160 446874 7

AF746400

PREFACE

Deep-water siliciclastic reservoirs are a major high-potential play. As of December 1994, more than 170 wildcats have been drilled in the deep-water Gulf of Mexico with thirty announced discoveries (Dirks and Van Den Berg, 1993) (Figure 1). Already more than 2.2 BOE have been discovered in deep water gravity-flow sands of Gulf of Mexico slope basins (Lawrence, 1994). The initiation of subsalt exploration adds importance to understanding deep-water sands, because most of the potential reservoirs will be of gravity-flow origin (McGuinness and Hossack, 1993). Additionally, recent large deep-water sand discoveries in the Paleogene of the North Sea and in the Tertiary of the western submarine slopes of Africa demonstrate the continued worldwide applicability of the turbidite play.

This SEPM Core Workshop 20 on Turbidites and Associated Deep-Water Facies has been assembled to examine the sediment and rocks deposited within eight slope basins of the Gulf of Mexico. One of the examples is Paleocene and seven are Plio-Pleistocene (Figure 2). The eight papers provide a starting point for discussions of depositional processes, facies, and reservoir and production characteristics based on observations from cores and core-photos. Turbidite is the most common word used to describe the sediment and rocks cored, but careful reading demonstrates that slumps, debris flows, high-density and low-density turbidites, and bottom-current reworked sediments are all recognized.

Geological observations of different data sets in exploration and other stratigraphic inquiry must be reconciled. Cores provide a high-resolution scale of observation but reflect limited areal and stratigraphic coverage. In contrast, seismic reflection profiles provide three-dimensional constraints on the depositional system, but are of low resolution, imaging packages of many individual events. Six of the authors in this volume integrate core photos and descriptions, seismic record sections and maps of depositional packages. These data sets are the focus of this SEPM Core Workshop and form the basis for evaluating different existing models and formulating new models for deposition of the deep-water Gulf of Mexico sands. Further discussions of turbidites and associated facies are available in the proceedings volume from the Gulf Coast Section SEPM Foundation Research Conference edited by Weimer et al. (1994).

REFERENCES

Lawrence, D.T., 1994, Turbidite Technical Challenges in the Deepwater Gulf of Mexico, *In*: Weimer, P., A.H. Bouma, and B.F. Perkins, (eds.), 1994, Submarine Fans and Turbidite Systems, Gulf Coast Section Society of Economic Paleontologists and Mineralogists Foundation, Fifteenth Annual Research Conference Proceedings, December 4-7, 1994, p. 217-220.

McGuinness, D.B., and J.R. Hossack, 1993, The Development of Allochthonous Salt Sheets as Controlled by the Rates of Extension, Sedimentation, and Salt Supply, *In:* Armentrout, J.M., R. Bloch, H.C. Olson, and B.F. Perkins, (eds.), Rates of Geologic Processes: Tectonics, Sedimentation, Eustasy, and Climate; Implications for Hydrocarbon Exploration: Gulf Coast Section Society of Economic Paleontologists and Mineralogists, Fourteenth Annual Research Conference Proceedings, December 5-8, 1993, p. 127-139.

Weimer, P., A.H. Bouma, and B.F. Perkins, (eds.), 1994, Submarine Fans and Turbidite Systems, Gulf Coast Section Society of Economic Paleontologists and Mineralogists Foundation, Fifteenth Annual Research Conference Proceedings, December 4-7, 1994, 440 p.

R.D. Winn, Jr. and J.M. Armentrout

WITHDRAWN

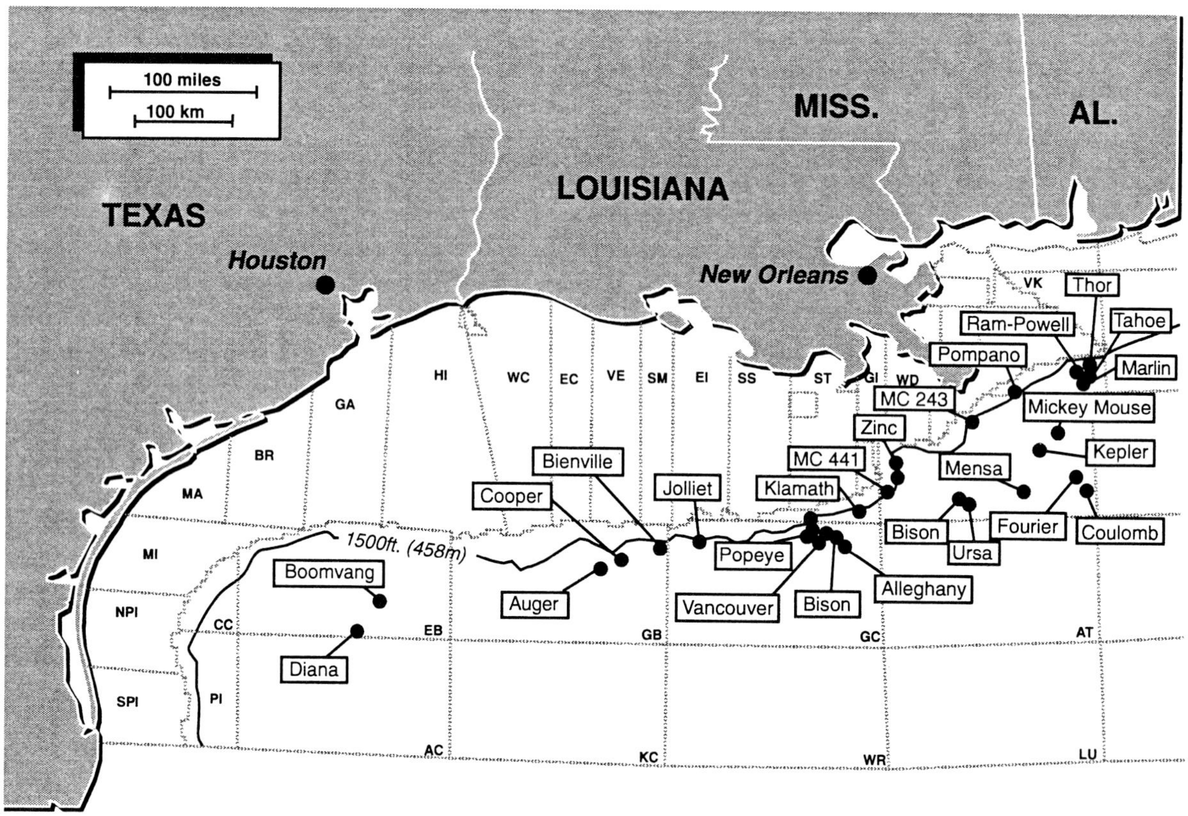

FIG. 1.—Announced turbidite discoveries in the deepwater Gulf of Mexico (Figure after Lawrence, 1994, Figure 1).

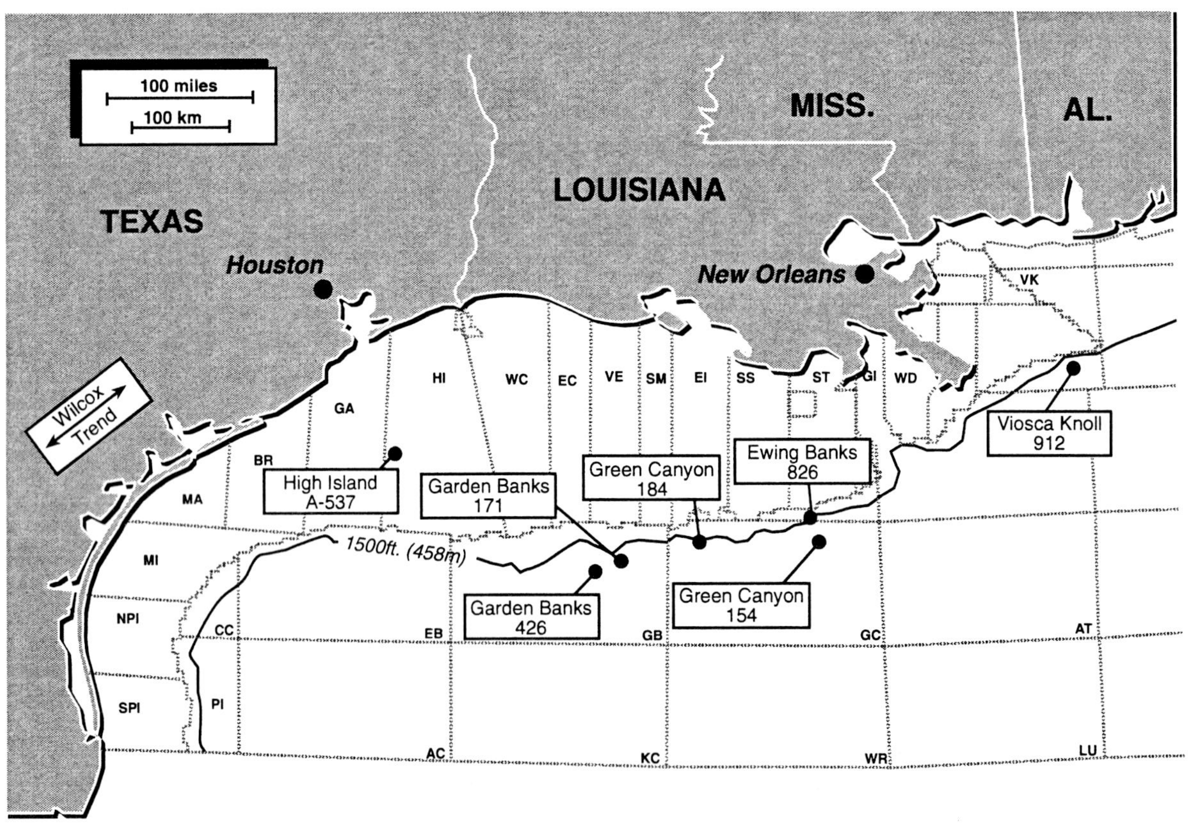

FIG. 2.—Location of core studies in this volume.

Turbidites And Associated Deep-Water Facies

TABLE OF CONTENTS

The SEPM gratefully acknowledges financial support from Mobil Oil Corporation for the publication of this volume.

Cover diagram provided by G. Shanmugam, Mobil Exploration and Producing Technical Center, Dallas, Texas

These SEPM Core Workshop Notes have received independent peer review. In order to facilitate rapid publication, these notes have not been subjected to the editorial review required for SEPM Special Publications.

Additional copies of this publication may be ordered from SEPM. Send your order to:

SEPM
P.O. Box 4756
Tulsa, Oklahoma 74159-0756
U.S.A.

© Copyright by SEPM (Society for Sedimentary Geology), 1995
ISBN 1-56576-018-2

Printed in the United States of America

TURBIDITES and ASSOCIATED DEEP-WATER FACIES

Organized and Edited

by

Robert D. Winn, Jr.
University of Papua New Guinea

and

John M. Armentrout
Mobil E&P Technical Center

1926 SEPM
Society for Sedimentary Geology

SEPM Core Workshop No. 20
Houston, Texas
March 5, 1995

© Copyright by SEPM (Society for Sedimentary Geology), 1995

552.5 S

CARDIFF
-2 JAN 1995

FACIES OF A SUBMARINE CANYON FILL RESERVOIR COMPLEX, LOWER WILCOX GROUP (PALEOCENE), CENTRAL TEXAS COASTAL PLAIN

WILLIAM E. GALLOWAY
THOMAS A. McGILVERY
Department of Geological Sciences
The University of Texas at Austin
Austin, TX 78712-1101

ABSTRACT

The Lavaca submarine canyon briefly formed a large slump-generated embayment, centered in Lavaca County, Texas, along the prograding Lower Wilcox shelf margin. Episodes of canyon excavation by large-scale mass wasting alternated with periods of delta lobe progradation into the upper canyon. Resultant canyon fill consists of (1) displaced, rotated slide blocks of the pre-canyon deltaic succession, (2) slump, debris flow, fluidized flow, and turbidity-flow deposits derived from canyon-margin deltas, which bypassed the steep (20 to 30°) canyon walls, and (3) progradational prodelta and delta front deposits. Slide, slump, and flow deposits onlapped the canyon floor; unstable delta front sediments infilled and healed the remnant canyon cut embayment. The lower, onlapping canyon fill contains typical mud-dominated turbidite channel fills and associated overbank and levee facies that are prolific, though elusive, gas-condensate reservoirs in Hallettsville E., Hallettsville Burns, Good Hope, and Campbell Creek fields.

The bulk of the canyon fill consists of mass-transport complexes of muddy slump and debris-flow lobes and tongues. Cohesive slides and slumps, a spectrum of liquefied flow deposits, high-density turbidites, and low-density turbidites are prominent in cores of the canyon fill. Reworked sand beds deposited by contour(?) currents are a minor but distinct lithofacies. In-situ hemipelagic and pelagic units are rare or absent. Reservoir facies consist of massive, amalgamated, coarse turbidite and grain flow units that are up to 100 ft (30 m) thick and confined to narrow, deep, leveed submarine channels. Interbedded mudflow units interrupt vertical continuity of sandstone within the channel fills. The trend and location of channel axes were controlled by the morphology of the enclosing canyon; the reservoirs form a series of offset, vertically superposed shoestrings within the muddy, onlapping, canyon fill. Adjacent deposits include heterolithic levee and overbank turbidites and current-reworked siltstone and fine sandstone. Levees were muddy, developed high relief, and locally collapsed as slides and slumps. Rotated slide blocks of pre-canyon delta-front sediment occur at the base of the canyon fill.

INTRODUCTION

In this paper we have three objectives. First, we describe the lithofacies of a Wilcox canyon-filling slope system of the NW Gulf of Mexico. Second, in so doing we illustrate a descriptive and interpretive scheme generally applicable to slope depositional systems. Finally, we place the descriptions and process interpretations into the overall context of slope and basin depositional systems. Slope and adjacent basin-floor sedimentation is dominated by gravity-transport processes, and the depositional record is commonly interrupted by local to regional erosion by the same processes. In addition, contour-parallel (geostrophic) and up- and down-slope marine currents may rework and remold slope deposits (Stow, 1986; Mutti, 1992; Shanmugam et al., 1993). This paper applies a genetic facies interpretation and classification scheme of Galloway and Hobday (in press) that identifies channel fills, turbidite lobes, sheet turbidites, low-density turbidite fills and sheets, slump and debris flow lobes and tongues, megaturbidite beds, hemipelagic drapes and fills, and contourite drifts as the main facies building blocks of slope and basin depositional systems.

Turbidite channels form the major framework facies within many slope and basin systems. They are scoured largely by high-density turbidity flows, but they may also be initiated, modified, or plugged by mudflows and debris flows. Channels may be erosional, depositional, or mixed (Fig. 1). Erosional channels typically occur on the steepest areas of the slope, where flows are accelerating. They display broad, V-shaped cross sections, marginal terrace scarps, or collapsed, cuspate banks. Depositional channel/levee complexes form where flows decelerate. They are broadly concave to flat-bottomed in cross section. Like fluvial channels, submarine channels form a spectrum from sand and gravel-dominated to mud-dominated types (Fig. 1). Coarse-grained channels are broad, low-relief, flat-based to broadly concave, and low sinuosity. With increasing mud content, levees form and channel stability and sinuosity increase. Submarine channels are large, commonly ranging from hundreds to thousands of feet in width and 100 to more than 300 ft (30 to 100 m) in depth. Together with their terminal *turbidite lobes*, levees, and adjacent overbank splays, turbidite channel systems form the most important framework elements of many slope systems. The channel/lobe complex displays predictable geometry and lithofacies associations, which are reflected in isolith pattern and log response (Fig. 2).

Slumps and debris flows form discrete sediment units that, in plan view, have sheet, lobe, and tongue geometries. An idealized slump exhibits several elements, including a proximal detachment scar, rotated and displaced blocks of slumped sediment, a lobe of plastically deformed cohesive sediment, and a distal tongue of liquefied sediment. In semi-solid slide blocks, original sediment bedding and textures are largely preserved, though deformed. With increasing liquefication and downslope flow of the sediment mass, original bedding and texture are lost and the resedimented debris or mud flows exhibit textural mixing, variable size grading, and crude bedding. Very large slump and debris-flow units are distinguished as mass-transport complexes.

Slope/basin depositional system classification has evolved from simple fan-based models to a much more realistic and flexible approach that emphasizes the pattern of sediment supply to the slope crest (point source vs. line source), the overall grain size of the supplied sediment, and the geomorphology of the resultant depositional system (Reading, 1991; Reading and Richards, 1994; Galloway and Hobday, in press). Distinct slope/basin system types include sand- and mud-rich fans, delta-fed aprons (or ramps), retrogradational aprons, submarine canyon fills, contourite drifts, and megaslide complexes. Within such slope/basin depositional systems, facies successions may be recognized and usefully subdivided using a high-resolution sequence stratigraphic framework. However, sequence stratigraphic models (e.g., Van Wagoner et al., 1990) place slope facies associations into a fixed, temporal progression that is inadequate for description and prediction of the contemporaneous paleogeographic (spatial) diversity of slope systems and facies clearly evidenced on modern basin margins. Consequently, we emphasize the facies-oriented paleogeographic approach here.

THE LAVACA SUBMARINE CANYON SYSTEM — SETTING, ORIGIN, AND HYDROCARBON PRODUCTION

The Lavaca canyon system was first identified by Chuber (1979). Further drilling of lower basal Wilcox deltaic reservoirs and the subjacent Edwards (Cretaceous) reef outlined two large, mud-filled erosional gorges incised into the progradational lower Wilcox delta system in Lavaca County, Texas (Chuber and Begeman, 1982). Discovery of commercial hydrocarbons within the canyon fill in 1983 focused exploratory drilling in the Hallettsville area. By 1989, 17 wells in 4 fields had produced about 900,000 bbl of oil and 10 Bcf of gas.

Chuber et al. (1989) and Galloway et al. (1991) utilized the extensive well and seismic data base created by exploration and development to interpret and describe the geometry, facies architecture, and origin of the Lavaca submarine canyon system and its contained reservoirs.

The Lavaca canyon system consists of the fill of at least three superimposed canyons, of which the Lavaca canyon is the youngest and most completely preserved. The canyons cut into outer-shelf and shelf-margin delta lobes of the lower Wilcox Rockdale delta system, along its southwestern flank (Fig. 3). Delta progradation in this area repeatedly advanced to the late Paleocene shelf margin, which lay above the lower Cretaceous shelf edge as defined by the Edwards reef, depositing the X, Y, Z, Svetlik, Technik, and Kubena parasequences (Fig. 4). Instability increased as the successive delta lobes advanced to and over the margin. Resedimented delta-front gravity flows and slumps are abundant, intraformational

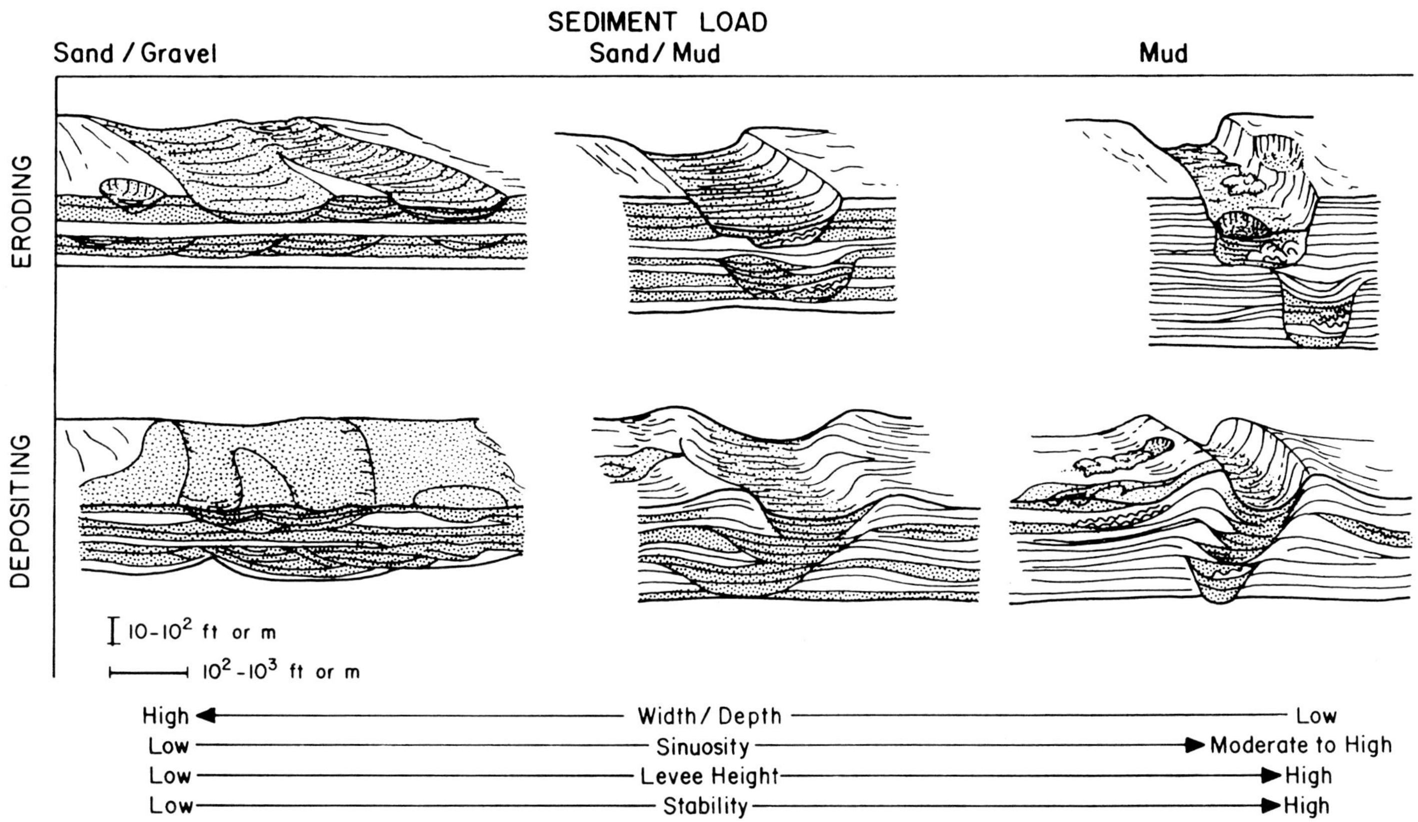

FIG. 1.—The spectrum of turbidite channels. Channels may be erosional or depositional; channel morphology depends on the ratio of sand to mud transported through the channel. From Galloway and Hobday (in press).

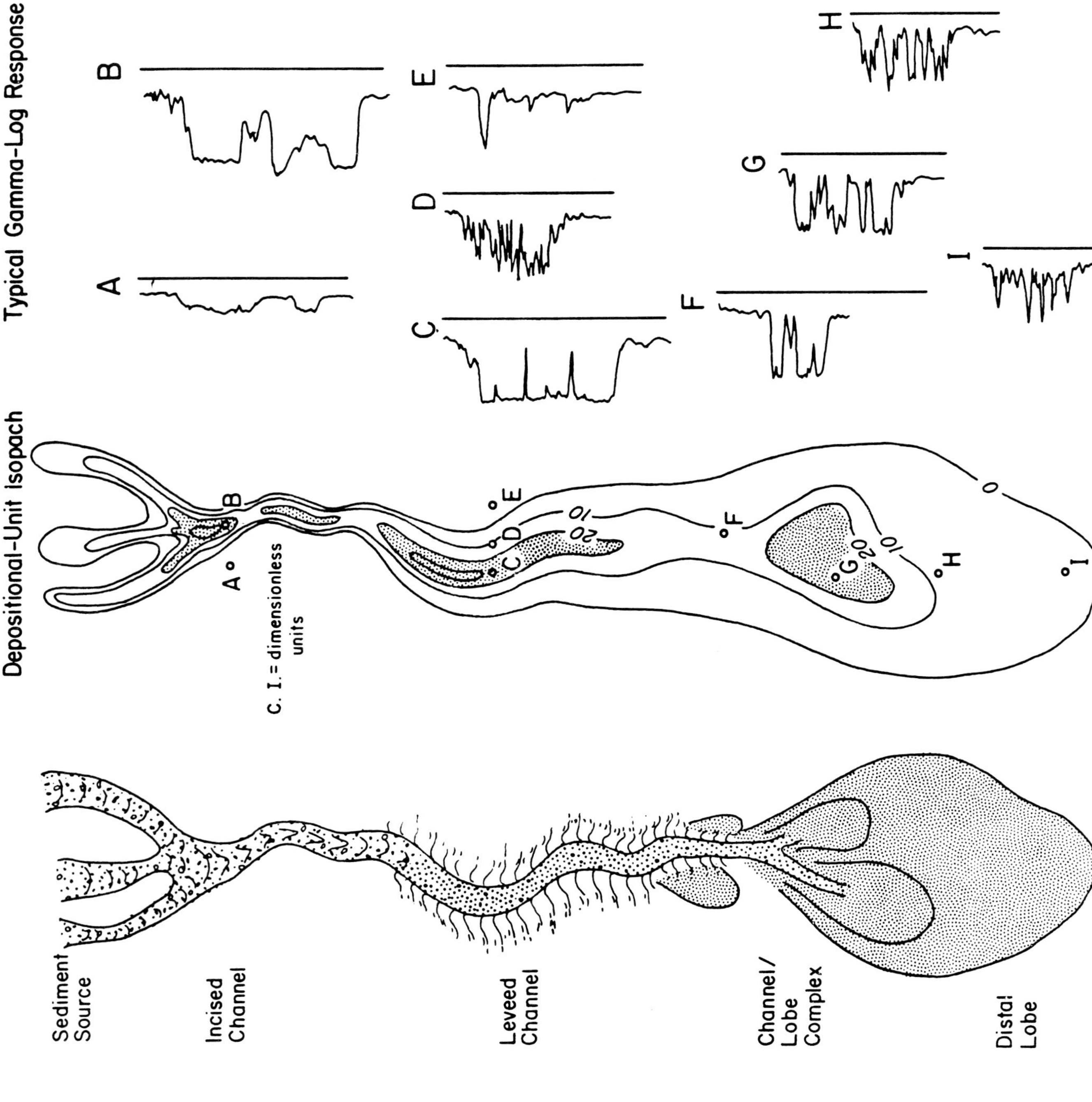

FIG. 2.—Environments, depositional geometry, and typical log response of a submarine channel/lobe complex. From Galloway and Hobday (in press).

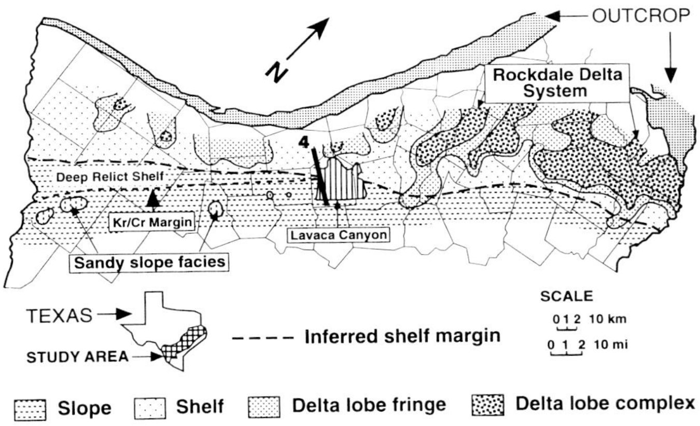

FIG. 3.—Paleogeographic location of the Lavaca canyon system on the flank of the lower Wilcox Rockdale delta system. Location of regional cross section (Figure 4) is on the SW flank of the Lavaca canyon. From Xue and Galloway (1993).

growth faults occur within the lower Wilcox and underlying Midway section, and the delta fronts are displaced along the first and most landward of the lower Wilcox growth fault zones.

Initial canyon excavation occurred following progradation of the Kubena parasequence (Fig. 4). The canyon was infilled by renewed delta progradation, reexcavated, again filled, and finally excavated and filled a third time. After the last canyon, the Lavaca, was filled, the lower Wilcox delta system renewed its progradation Gulfward onto the continental slope. Remnants of the earlier canyons are preserved only where they project from beneath the flanks of the Lavaca canyon. The morphology of the Lavaca canyon has been reconstructed by decompacting an isopach of the canyon fill, which provides a cast of the canyon morphology (Galloway et al., 1991). The canyon was nearly 4,000 ft deep where it opened out onto the lower slope and formed a broad, funnel-shaped embayment in the lower Wilcox continental margin (Fig. 5). Reconstructed canyon wall slopes were 20° to 30°, and arcuate slump scars outlined the mapped canyon margin. Internally, the canyon thalweg was terraced, reflecting the listric geometry of the buried and preserved slump planes left by the largest slope failures. Canyon morphology clearly reflects its excavation by retrogressive slumping of the upper slope and shelf margin.

Palynomorphs and foraminifera extracted from the Golsch #4 and Allen #2 cores of the canyon fill reveal a sparse, environmentally restricted, and nondiagnostic assemblage (Chuber et al., 1989). Reworked Cretaceous spores and pollen are common, especially in the Golsch core samples. Foraminifera indicate a restricted marine setting for the muddy canyon fill, and a shallow marine setting for an interpreted displaced deltaic slide block in the lower Allen #2 core.

The origin of the Lavaca canyon system has been interpreted by Galloway et al. (1991) using the model developed by marine geologists for formation of modern canyons on depositional continental margins (Farre et al., 1983; Carlson and Karl, 1988). Progradation of earliest Wilcox delta lobes onto the paleoshelf margin loaded and destabilized underlying, muddy, slope deposits. Slumping of the unstable margin, in turn, excavated a broad embayment in the upper slope and shelf margin. Failure was likely localized geographically by the reactivation of a buried upper Cretaceous megaslide scar, which directly underlies the Lavaca canyon system (Berg, 1979). Retrogressive slumping and mass wasting of the headwall of the slump scar expanded the proto-canyon onto the shelf and deposited an onlapping debris apron at the base of the slope. As subsequent delta lobes praded onto the steep, high-relief canyon walls, debris and turbidity flows spilled down the canyon, creating marine density currents that scoured canyon tributaries and deposited a thick, onlapping, lower canyon fill. The turbulent fluid flows developed organized channel/levee complexes with a well-defined differentiation of sandy and muddy facies. Sediment supply was mud rich, except at active distributary mouths where sand was funneled directly into the canyon. As the canyon back-filled, gradients and differential relief were reduced, and relatively stable prodelta slopes were able to prograde into the remnant canyon. Channel-mouth bypassing and gravity remobilization continued to supply gravity flows, which formed debris-flow lobes and turbidite channel fills.

The canyon fill is dominantly mudstone. Sandstone bodies are discontinuous and unpredictable in distribution (Fig. 4). However, the overall occurrence of sandstone reflects the influence of the bounding canyon morphology. Geographically, sandstone facies occur along the canyon thalweg and in the middle canyon area (Fig. 5). Stratigraphically, sandstone facies occur within both the lower onlap and overlying offlap canyon fill intervals. Sandstone units within the offlap section can be traced updip to superjacent, progradational, deltaic sand facies (Galloway et al., 1991, their Fig. 13.16), whereas sandstone facies within the onlap fill are mapped as isolated

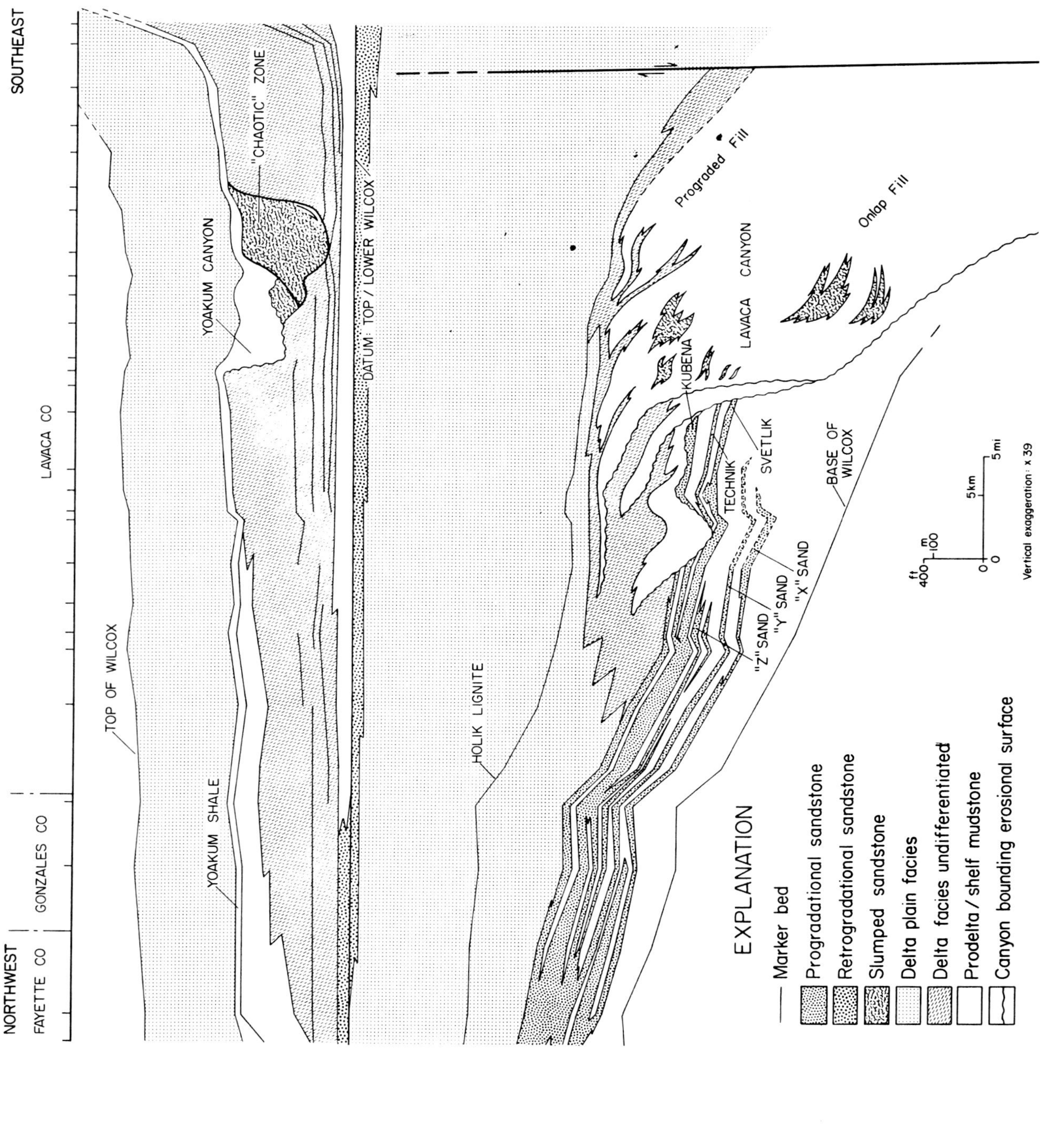

FIG. 4—Dip cross section across the lower Wilcox continental margin and across the southwestern flank of the Lavaca canyon. Section datum is the middle Wilcox (Yoakum shale) flooding surface. The Lavaca canyon consists of three nested canyon cut surfaces, the oldest of which formed soon after progradation of the Kubena sandstone parasequence. Cores described here are all from the lower onlap fill. For location, see Figure 3. From Galloway et al. (1991).

pods and narrow, sinuous shoestrings that trend along the canyon axis (Galloway et al., 1991, their Fig. 13.15; Chuber et al., 1989).

The canyon-cut surfaces would constitute sequence boundaries for depositional sequences (van Wagoner et al., 1990). In the depositional sequence model, the onlapping canyon fill facies that are the focus of this core workshop comprise the slope-fan complex of the lowstand systems tract. Offlap deposits that cap the canyon fill constitute the lowstand prograding wedge.

The process facies and genetic facies associations of the Lavaca canyon fill are typical of the slope sediments of the Gulf Coast Cenozoic. The fill records deposition by mudflows, debris flows, and turbidity currents on a dominantly progradational, shelf-edge delta-fed slope. Framework facies are turbidite channel and channel/lobe complexes, which are encased within muddy debris flow, low-density turbidite, and hemipelagic sediment. The distribution and geometry of the framework sandstone facies are strongly constrained by the morphology of the bounding canyon-cut unconformity (Fig. 5). The genetic association of canyon excavation and coherent failure by sliding and slumping is reflected in the preservation of large slide blocks and slump units containing displaced deltaic sands.

FACIES DESCRIPTION AND INTERPRETATION

Turbidite and associated facies of the Lavaca canyon fill are illustrated by seven cores from five wells drilled in the Hallettsville area (well locations shown in Fig. 6). Cores are from comparable depth and approximate strati-

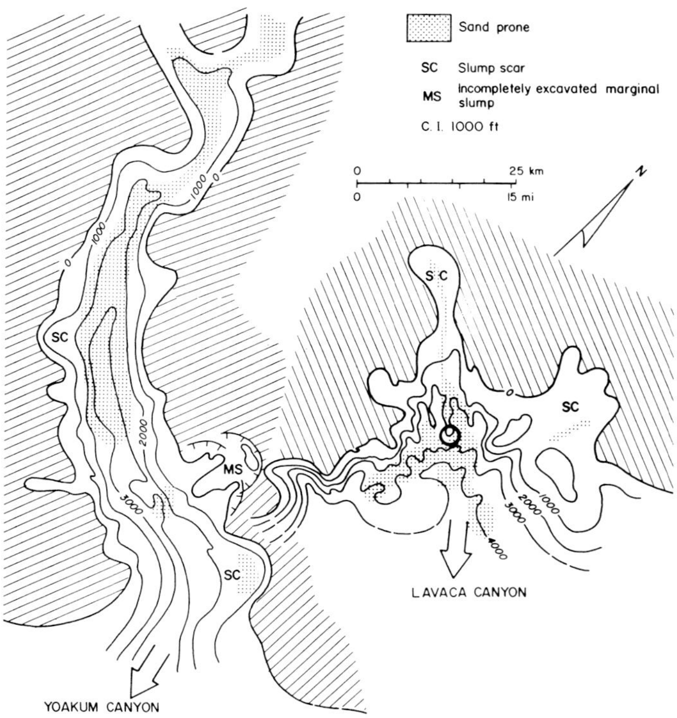

FIG. 5.—Geometry of the Lavaca and younger, middle Wilcox Yoakum canyons based on decompacted isopach maps of their fills. Note localization of sand-prone facies in the canyon axes and adjacent to tributaries. The canyons have been offset along strike to prevent overlap of contours. Circle locates Hallettsville area (Fig. 6). Modified from Galloway and Hobday (in press).

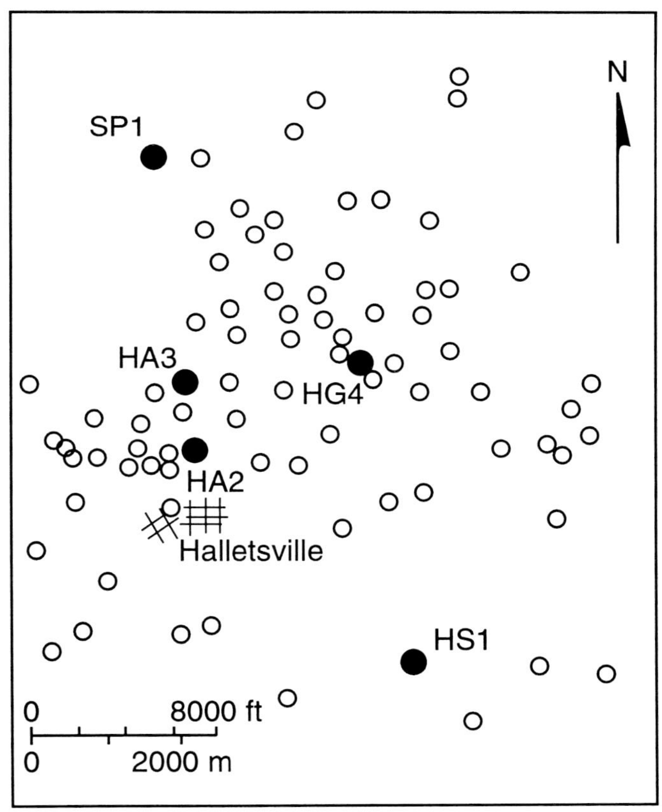

FIG. 6.—Location map for wells in the Hallettsville area (Fig. 5). Cored wells are indicated by the black dots. SP1; Speedwill O & G Pavliska #1; HA3, Howell Drilling Allen #3; HG4, Howell Drilling Golsch #4; HA2, Howell Drilling Allen #2; HS1, Howell Drilling Speedwell #1.

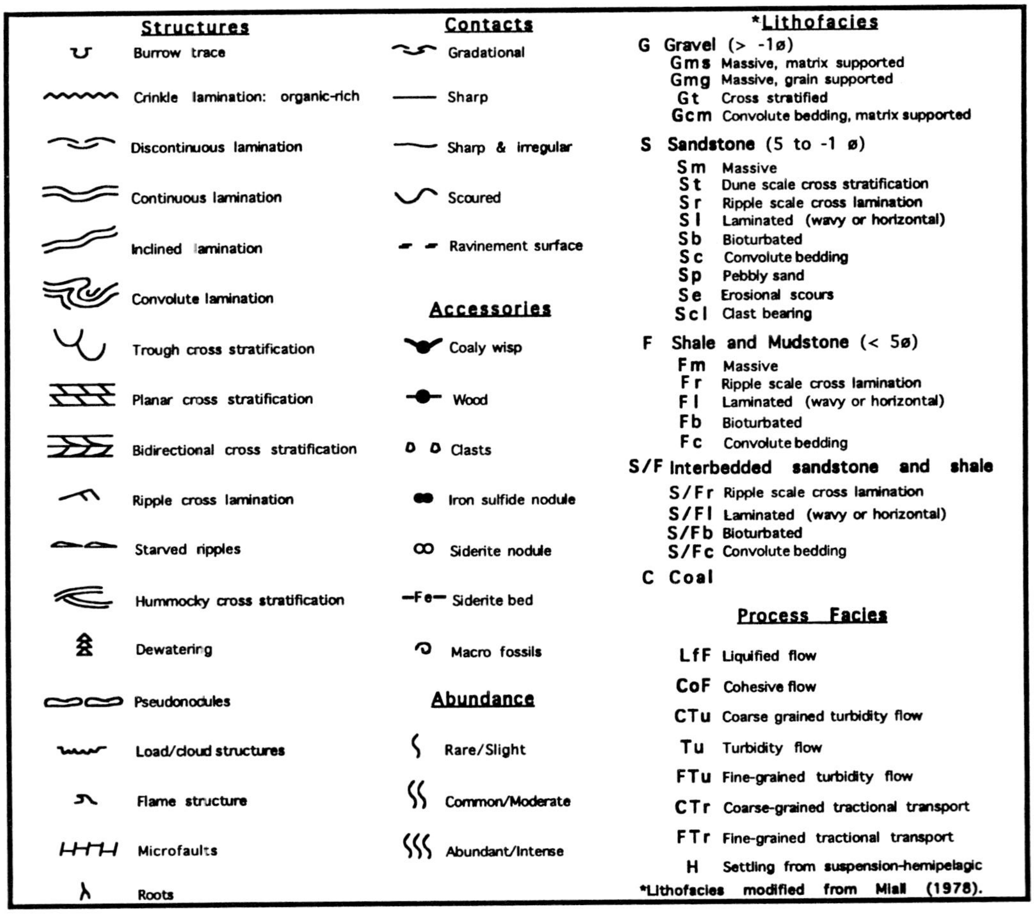

FIG. 7.—Explanation of symbols and codes used on graphical core description forms.

graphic position within the canyon fill. Represented intervals include the Hathaway and Golsch zones within the Lavaca canyon fill and unnamed intervals in the underlying Smothers canyon fill (Chuber et al., 1989). However, exact correlations among the facies cored are uncertain because of the depositional complexity and intraformational erosional and structural discontinuities that dominate unstable slope successions. Graphical descriptions (Figs. 7-12) of six representative core intervals summarize observations and interpretations. Additional well information includes selected thin section, porosity, permeability, and paleontological data (Chuber et al., 1989), and well log suites (Fig. 13). Detailed subsurface lithofacies maps and regional well and seismic cross sections are found in Paige (1988), Chuber et al. (1989), and Galloway et al. (1991).

First, the lithology, sedimentary structures, and accessory components were graphically logged. Principal *lithofacies* units and their contacts were described and coded using a scheme (Fig. 7) modified from that of Miall (1978). Descriptive information was then used to interpret *depositional process* facies, as defined at the lithofacies unit or bed scale (Fig. 7). The principal process facies that compose slope and basin systems are (1) cohesive slide and slump units, (2) liquefied deposits, including mudflow, debris flow, and fluidized flow beds, (3) high-density turbidites, (4) low-density turbidites, (5) contourites, (6) tractionites, (7) hemipelagites, and (8) pelagic drapes (Stow, 1986; Pickering et al., 1989; Mutti, 1992). Together, process facies are organized into a number of genetic facies that reflect both the depositional process and the specific environment. These include (1) turbidite channel fills (and associated levee deposits), (2) turbidite lobes, (3) sheet turbidites, (4) slump and debris-flow lobes and tongues, (4) low-density turbidite fills and sheets, (5) mass-transport complexes (also called megaturbidites), (6) contourite drifts, and (7) hemipelagic fills and drapes.

Depositional facies associations illustrated by the cores include (Figs. 13 and 14):

1. Turbidite channel-fill sandstone and subordinate mudstone (A).
2. Channel-margin levee and overbank interbedded sandstone, shale, and laminite (thinly interbedded to interlaminated sandstone and shale). Channel-margin facies include levee (B) and isolated splay turbidite beds (C).
3. Unconfined mass-transport deposits consisting of slump and debris flow mudstone, sandy mudstone, and mudstone (D).
4. Interchannel and reworked levee deposits including contour(?)-current reworked fine sandstone (E). In addition, intact slide blocks of displaced levee (F) and delta front (G) facies form a significant component of the canyon fill.

The environmental setting and spatial organization of facies are shown schematically in Figure 14. Thickest sandstone bodies consist of amalgamated turbidites that filled muddy, leveed aggradational submarine channels. Channel fill units are flanked and locally underlain or overlain by thin to medium, interbedded sandstone and shale deposited in channel-margin levee and overbank settings. Detailed maps and log cross sections based on closely spaced wells (Chuber et al., 1989) show that the Lavaca channel fills are narrow (<1,000 ft or 300 m), lenticular shoestring sand bodies bounded by prominent, heterolithic levee deposits, as predicted by the channel spectrum (Fig. 1). The channel/levee complexes form the coarse-grained skeletal framework of the canyon fill and are the principal reservoir facies. Closely associated with the channel-margins are turbidity- and contour-current deposited sandstone beds. High depositional relief and rapid aggradation caused levee flanks to fail and slump both into the channel and onto the adjacent muddy interchannel areas. Interchannel areas contain mass-transport units composed of muddy to heterolithic, cohesive to fully liquefied sediment lobes and tongues. Isolated slide blocks of delta front sediment occur in the oldest canyon fill, and are likely remnants of the excavation phase that remain perched high on the landward canyon wall.

Turbidite Channel Fill Facies Association

Core intervals: Figure 8, Golsch #4, 9903-10,088 ft
Figure 9, Speedwell #1, 9946-9975 ft
Core photographs: Figure 15, A-E; Figure 16, A-C
Lithofacies composition: The dominant lithofacies include massive fine- to medium-grained sandstone (Sm) and mud-clast-bearing sandstone (Scl), particularly at the base of the fill unit. Interbedded laminated sandstone (Sl), massive mudstone (Fm), laminated mudstone (Fl), and convolute mudstone (Fc) are typical. Ripple- and convolute-laminated sandstones (Sr, Sc) also occur in the upper channel fill. Bedding contacts are sharp and commonly scoured.
Structures and accessories: Massive, structureless beds dominate. Additional structures include wavy, parallel, convolute, and ripple lamination and dewatering structures. Disseminated carbonaceous debris, mudstone rip-up clasts, and reworked siderite nodules, as well as in-situ early diagenetic siderite nodules and bands, are common accessory constituents.
Depositional processes: Coarse-grained, high-density turbidity flows and sandy to muddy, liquefied flows dominated channel deposition. Flow types became more diverse late in the history of channel filling; cohesive-flow and fine-grained turbidity-flow units are contained in the upper fill, and a discrete block of slumped, contorted levee sediment occurs at 9942-9951 ft in the Golsch core (Fig. 15D). Flows were viscous to concentrated, erosive, and confined within a narrow channel that is an excellent example of the muddy turbidite channel (Fig. 1).
Comments: Channels were likely nucleated as a slump or mudflow chute. The basal unit in the Golsch core (Fig. 8) is a contorted, hackley-fractured mudstone interpreted to be a dense, highly viscous mudflow deposit. While active, the channel filled with amalgamated coarse-grained turbidites (Fig. 15A). As abandonment occurred, the channel filled with a diverse mixture of coarse-grained turbidites, mud flows, sandy debris flows, and locally sourced channel-bank slump and slide debris (Fig. 15B, C, and E). The result is a crudely upward-fining log signature (Fig. 13). The cored channel fill is capped by interbedded thin sandstone and shale (laminite) of a levee. Sandy turbidite channel fill is the principal reservoir facies. Typical porosity is 21 to 22 percent; permeabilities range from 10 to 50 md (Chuber et al., 1989).

Channel Margin Levee and Overbank Facies Association

Core intervals: Figure 8, Golsch #4, 9881-9903 ft
Figure 9, Speedwell #1, 9975-10,023 ft
Core photographs: Figure 16 D, E

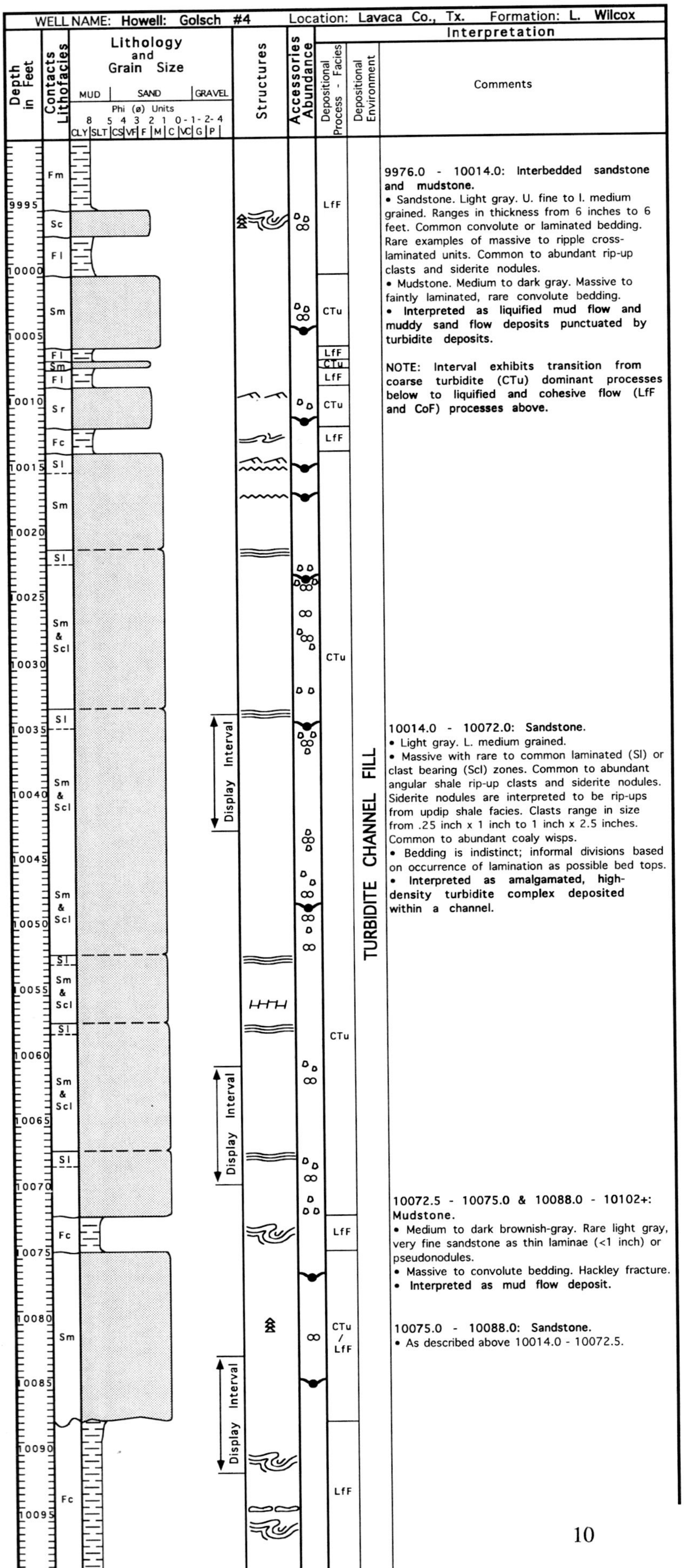

FIG. 8A.—Graphical core description and interpretation, Howell Drilling Golsch #4.

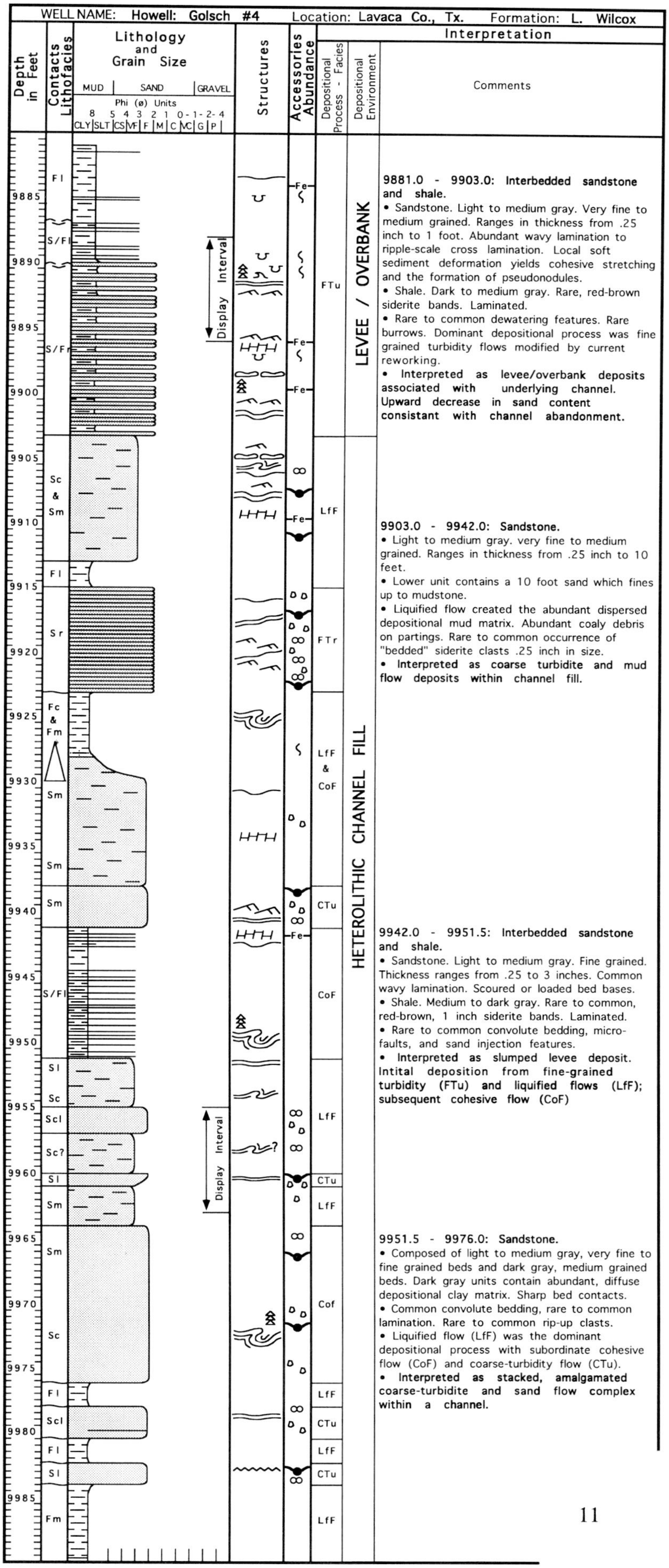

FIG. 8B.—Graphical core description and interpretation, Howell Drilling Golsch #4.

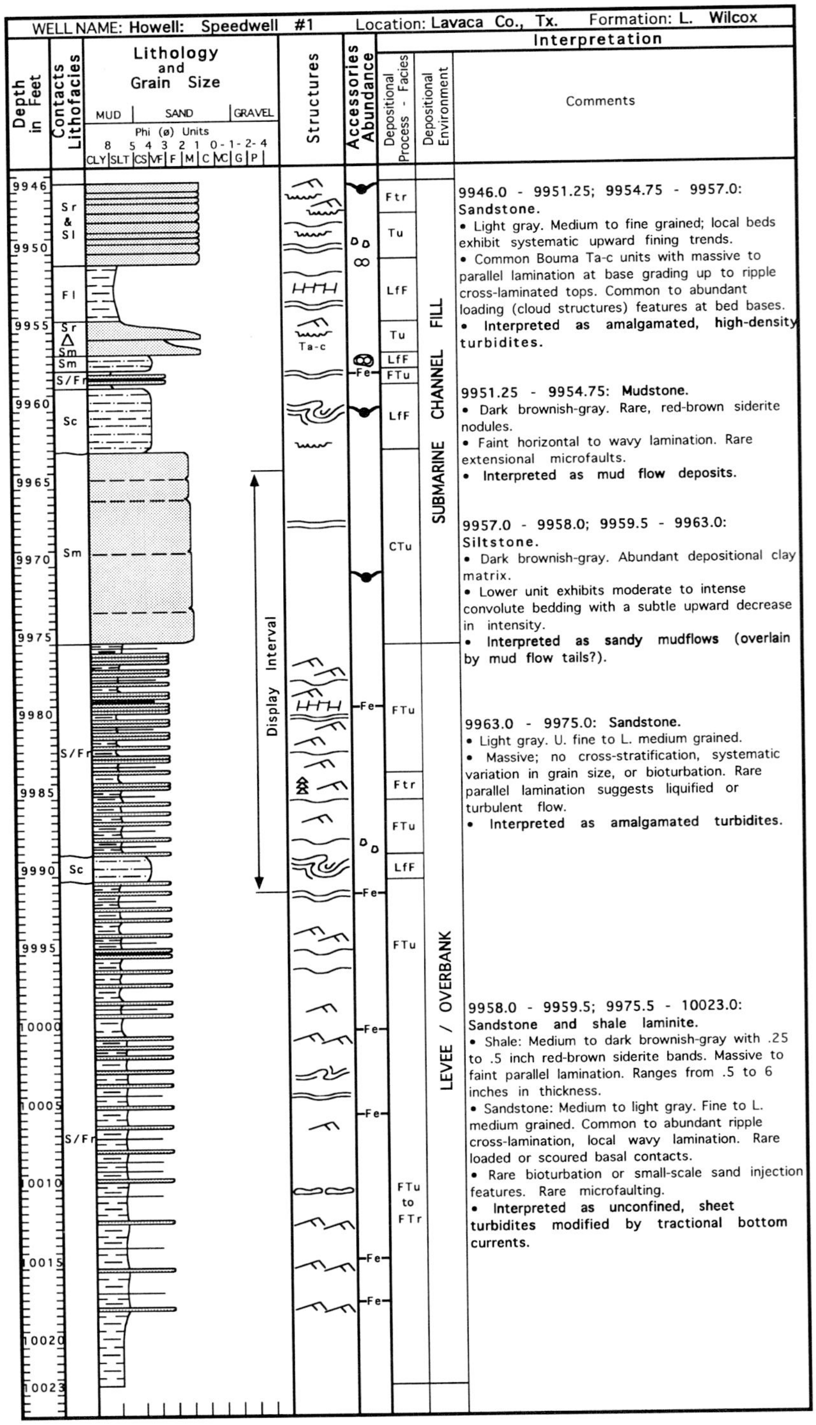

FIG. 9.—Graphical core description and interpretation, Howell Drilling Speedwell #1.

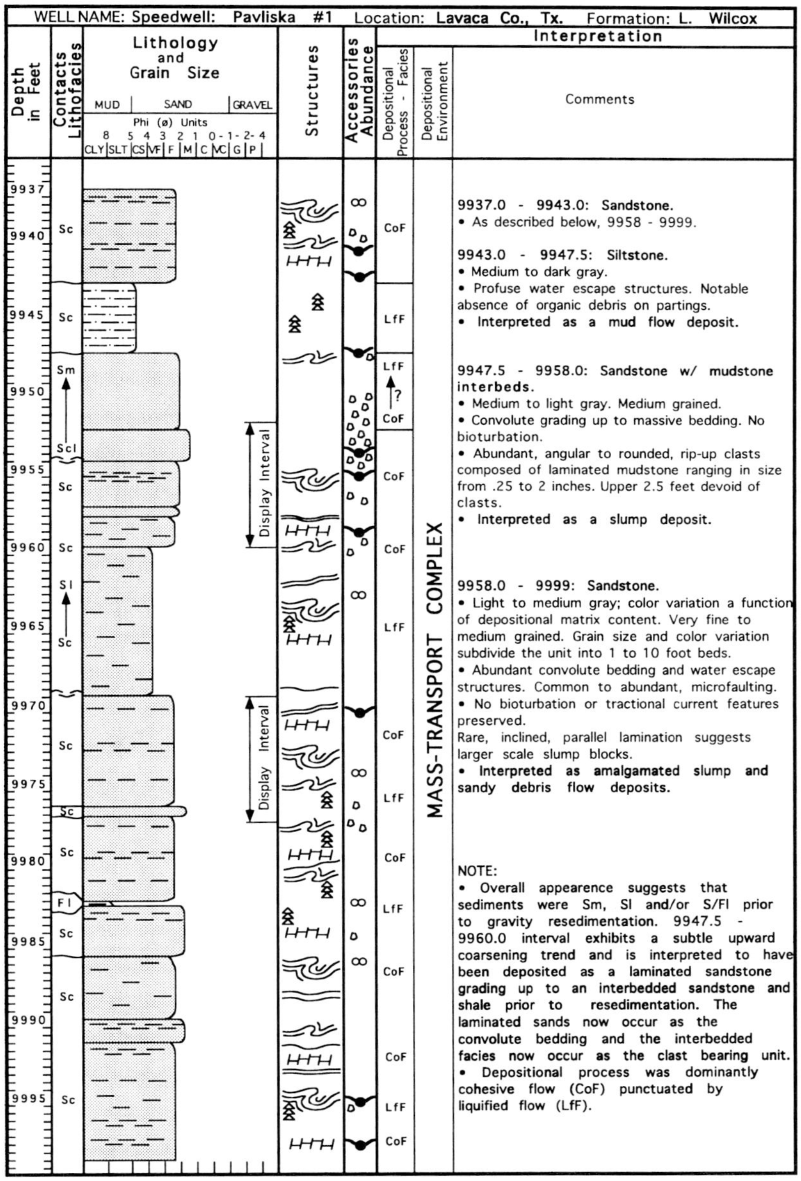

FIG. 10.—Graphical core description and interpretation, Speedwell O&G Pavliska #1.

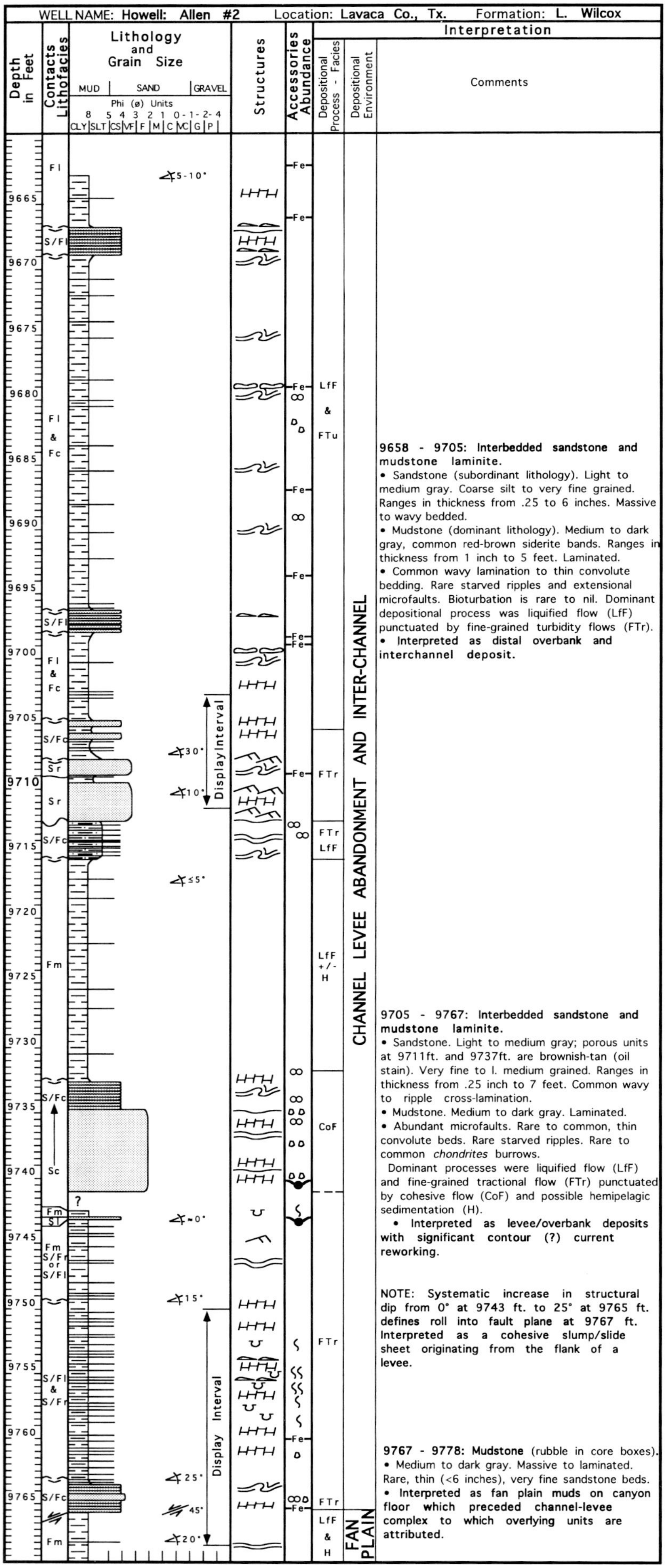

FIG. 11.—Graphical core description and interpretation, shallow interval of the Howell Drilling Allen #2.

Lithofacies composition: Thinly interbedded, very fine to fine sandstone and shale and laminites displaying wavy, lenticular, and ripple lamination (S/Fl, S/Fr) dominate the facies association. Laminated shale (Fl), convolute beds (S/Fc), and thin to medium beds of ripple-laminated sandstone (Sr) are present.
Structures and accessories: Microfaults, dewatering structures, sparse bioturbation, load casts, pseudonodules, and diagenetic sideritic bands are common.
Depositional processes: Deposition was dominated by fine-grained, low-density turbidity flows and low-energy traction-current reworking. Minor slumping, extension, and loading locally deformed or resedimented levee deposits.
Comments: Levee and proximal overbank deposits occur adjacent to depositional turbidite channel fills. With channel migration, levee facies also underlie and overlie channel-fill facies, as seen in the Golsch and Speedwell cores.

Mass-Transport Unit

Core intervals: Figure 10, Pavliska #1, 9937-9999 ft
Core photographs: Figure 15 F; Figure 16 F, G
Lithofacies composition: Dominantly muddy, very fine to medium, massive, convolute sandstone units (Sc) show sharp bases and are interbedded with laminated sandstone and mudstone (Sl, Fl).
Structures and accessories: Rounded to angular lithoclasts (Fig. 16 G) are common in some intervals. Swirled, diffused bedding is typical (Fig. 16 F). Dewatering structures, microfaults, carbonaceous plant debris, and diffuse siderite nodules are characteristic features. Steep dips, diffused or shattered relict textural bedding, and poor sorting are prominent in the core.
Depositional processes: Sediments were deposited by a spectrum of gravity mass flows ranging from cohesive to highly liquid.
Comments: Mapping of the canyon fill (Galloway et al., 1991, their Fig. 13.15) shows sandy mass-transport units to have irregular tongue to lobe geometries.

Reworked Interchannel and Levee Facies Association

Core intervals: Figure 11, Allen #2, 9663-9767 ft
Core photographs: Figure 16 H
Lithofacies composition: The heterolithic interval is composed of wavy to convolute shale (Fl, Fc), massive mudstone (Fm), and ripple, wavy, and convolute laminated very fine to fine sandstone (Sr, Sl, Sc) and laminite (S/Fr, S/Fl, S/Fc).
Structures and accessories: Microfaulting in sandy intervals, scattered bioturbation (*chondrites)*, and diagenetic siderite nodules and bands are common.
Depositional processes: An interplay of fine-grained, low-density turbidity flows and traction currents produced laminated beds; mudflows deposited massive and convolute intervals. Discrete, well-winnowed, and finely laminated sand beds at 9711 and 9735 ft are interpreted as contourites consisting of sand reworked from channel fill or levee turbidites. Variable dips suggest creep-induced extension and deformation reflecting rapid deposition on a sloping substrate. The progressive downward dip increase and the sharp, basal 45° glide plane displayed by the laminite unit at 9743-9765 ft show it to be a rotated slide block of levee facies. The upward progression from slumping to current reworking and finally to muddy deposition suggests abandonment or migration of the adjacent channel.
Comments: Contourite sand beds, although thin, are permeable and oil stained. As deposits reworked from an adjacent channel/levee complex, they form a marginal sandy halo flanking the channel belt. The siderite nodules and bands that are abundant in the interchannel mudstone are likely sources of the siderite clasts reworked into channel fills.

Coherent Delta-Front Slide Block

Core intervals: Figure 12, Allen #2, 9922-10,058 ft
Core photographs: Figure 15 G through I; Figure 16 I
Lithofacies composition: The section consists of massive, fine- to medium-grained sandstone units (Sm) capping upward-coarsening successions of interbedded sandstone and shale (Fig. 13) displaying variable mixtures of ripple, wavy, and convolute lamination and bioturbation (S/Fr, S/Fl, S/Fc, S/Fb). Massive, muddy sandstone beds (Sm, Fm) also occur.
Structures and accessories: A diverse and abundant ichnofauna, including *Ophiomorpha* (Fig. 16 I), a common shoreface form, distinguishes these laminites from those in cores of the canyon fill. Carbonaceous debris, microfaults, dewatering structures, and siderite nodules and bands are common.
Depositional processes: The deep core from the Allen #2 consists of a unique, displaced but coherent slide block of shelf-margin deltaic facies deposited before canyon

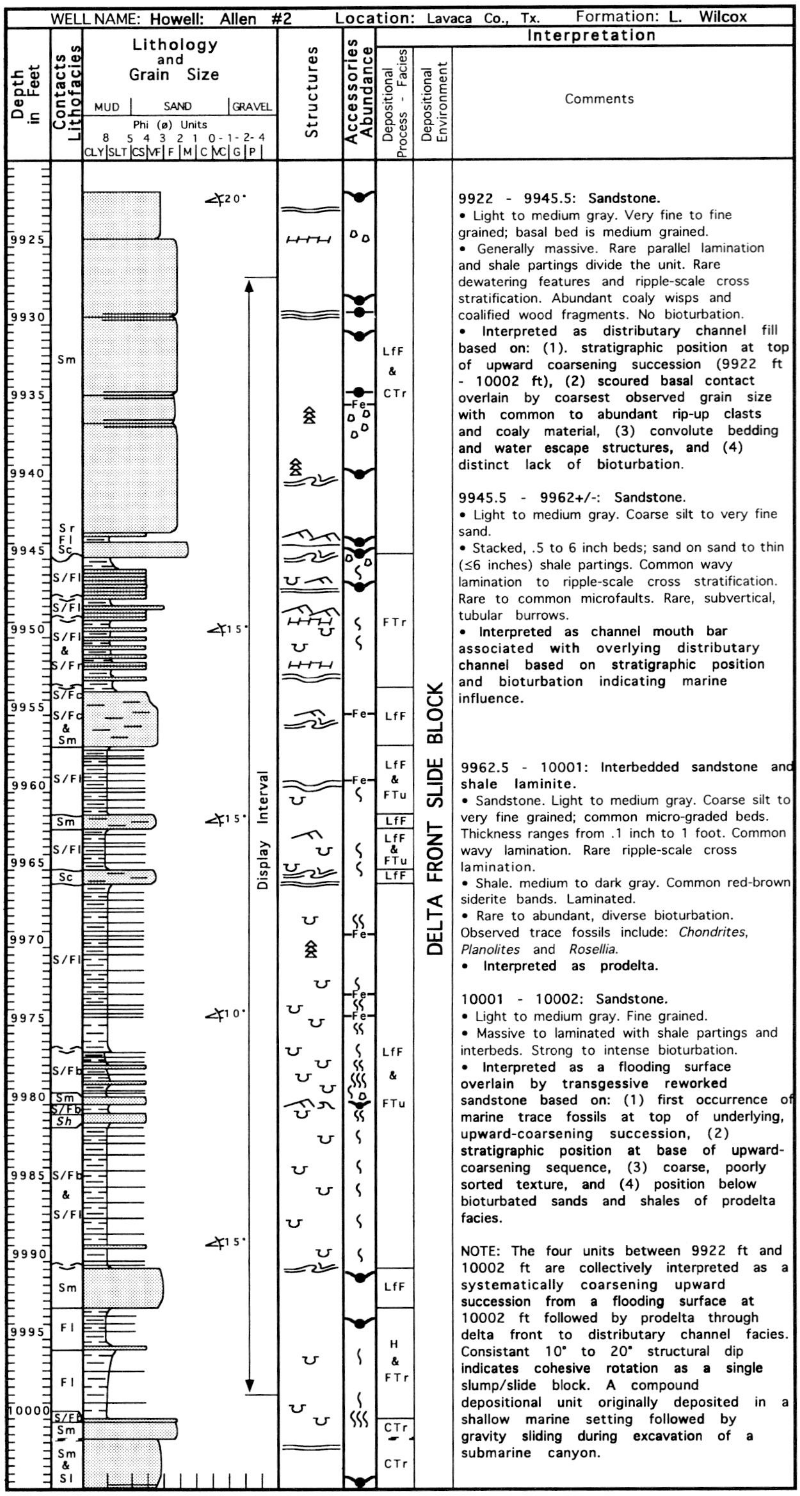

FIG. 12A.—Graphical core description and interpretation, deep interval of the Howell Drilling Allen #2.

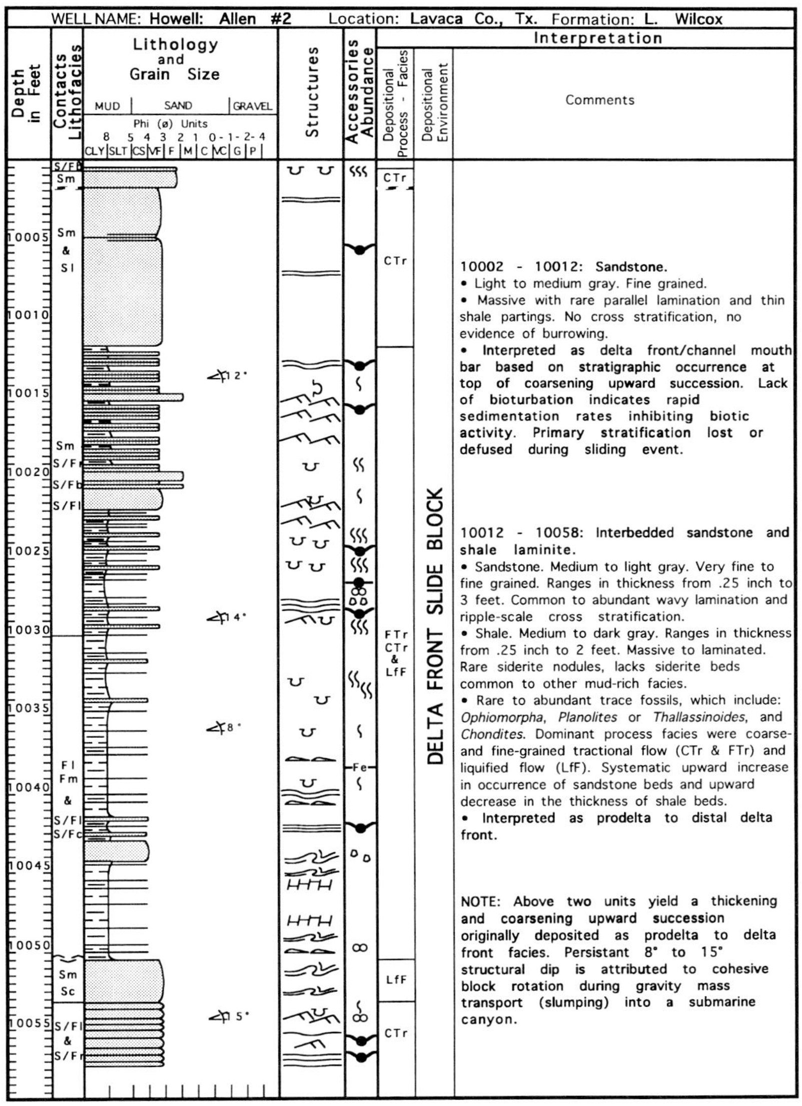

FIG. 12B.—Graphical core description and interpretation, deep interval of the Howell Drilling Allen #2.

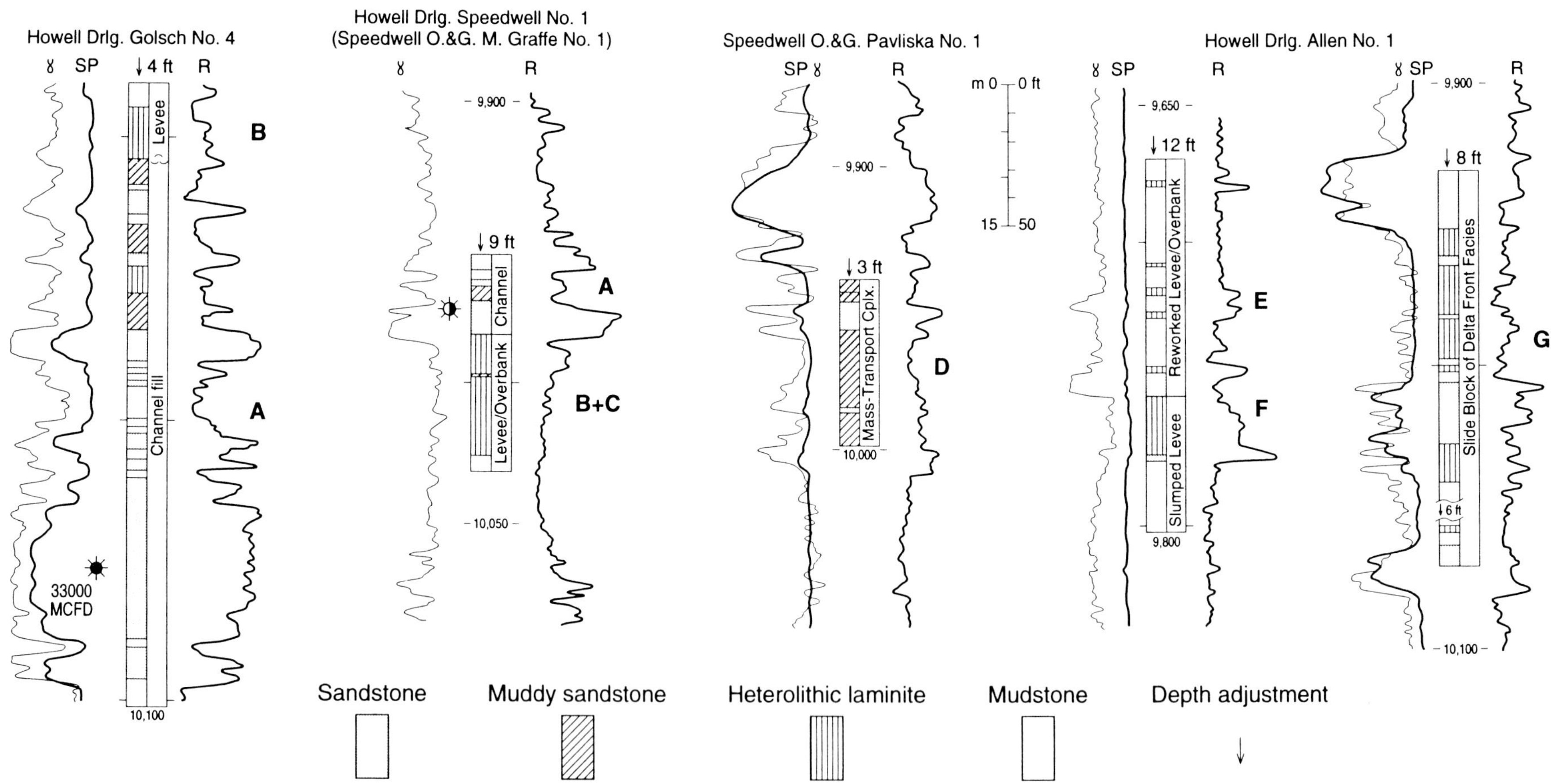

FIG. 13.—Electric and gamma-ray logs of the cored intervals discussed here. Inset within each log shows the dominant lithology and interpreted facies association. Core depths have been adjusted to match the log depth. Capital letters locate facies associations discussed in the text and in Figure 14. Core intervals are from several closely associated zones within the Lavaca and subjacent Smothers canyon fills.

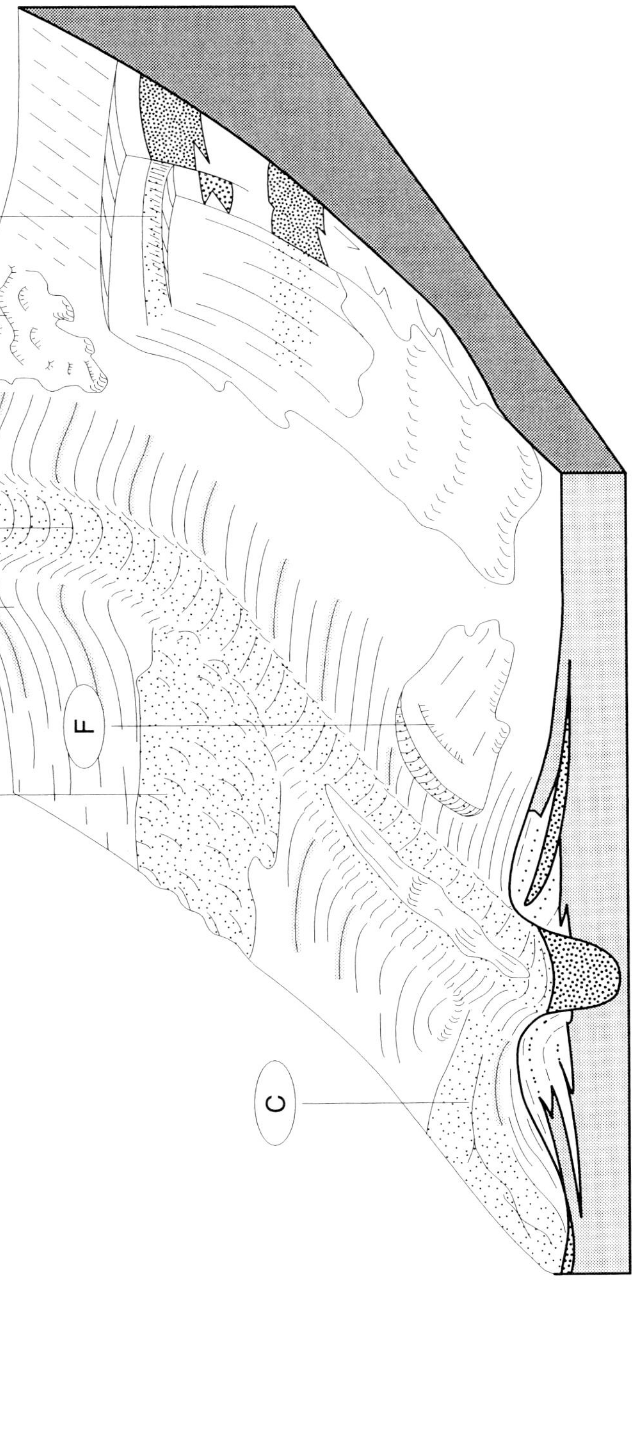

FIG. 14.—Schematic diagram of the environments and depositional features illustrated by the Hallettsville core suite. Letters match those adjacent to logs shown in Figure 13.

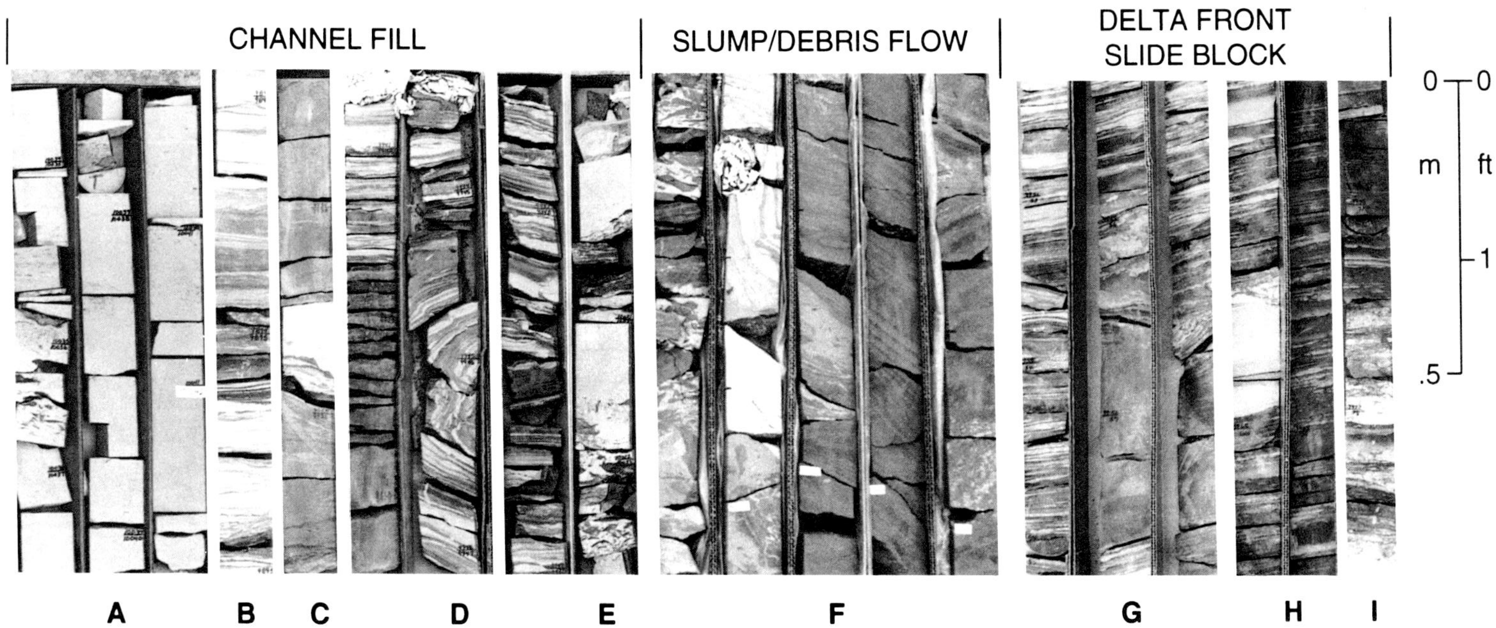

FIG. 15.—Selected whole-core segments. A—E: Representative lithofacies of the turbidite channel fill (Howell Drilling Golsch #4). From base to top, the fill includes amalgamated, massive high-density turbidites (A) with floating mud clasts and carbonaceous fragments, interbedded current-rippled sandstone, thin, low-density turbidites, and mudflow beds (B), liquified sand and mudflow deposits (C), recumbently folded slump unit of levee sediment (D), and sandy debris flows with abundant, locally scoured mud clasts (D). F: Representative 15-ft section of a mass-transport unit (Speedwell O&G Pavliska #1). G-I: Representative lithofacies of the prodelta and distal delta front facies contained in the lower Howell Allen #2 core. Ripple and wavy laminites with interbedded, massive mudflow beds (G) grade downward into shale containing sparse sandy frontal-splay turbidite beds (H). Abundance and diversity of bioturbation (I) distinguish the prodelta laminites from those of the deep-water, restricted canyon fill. Note the pervasive dip throughout core.

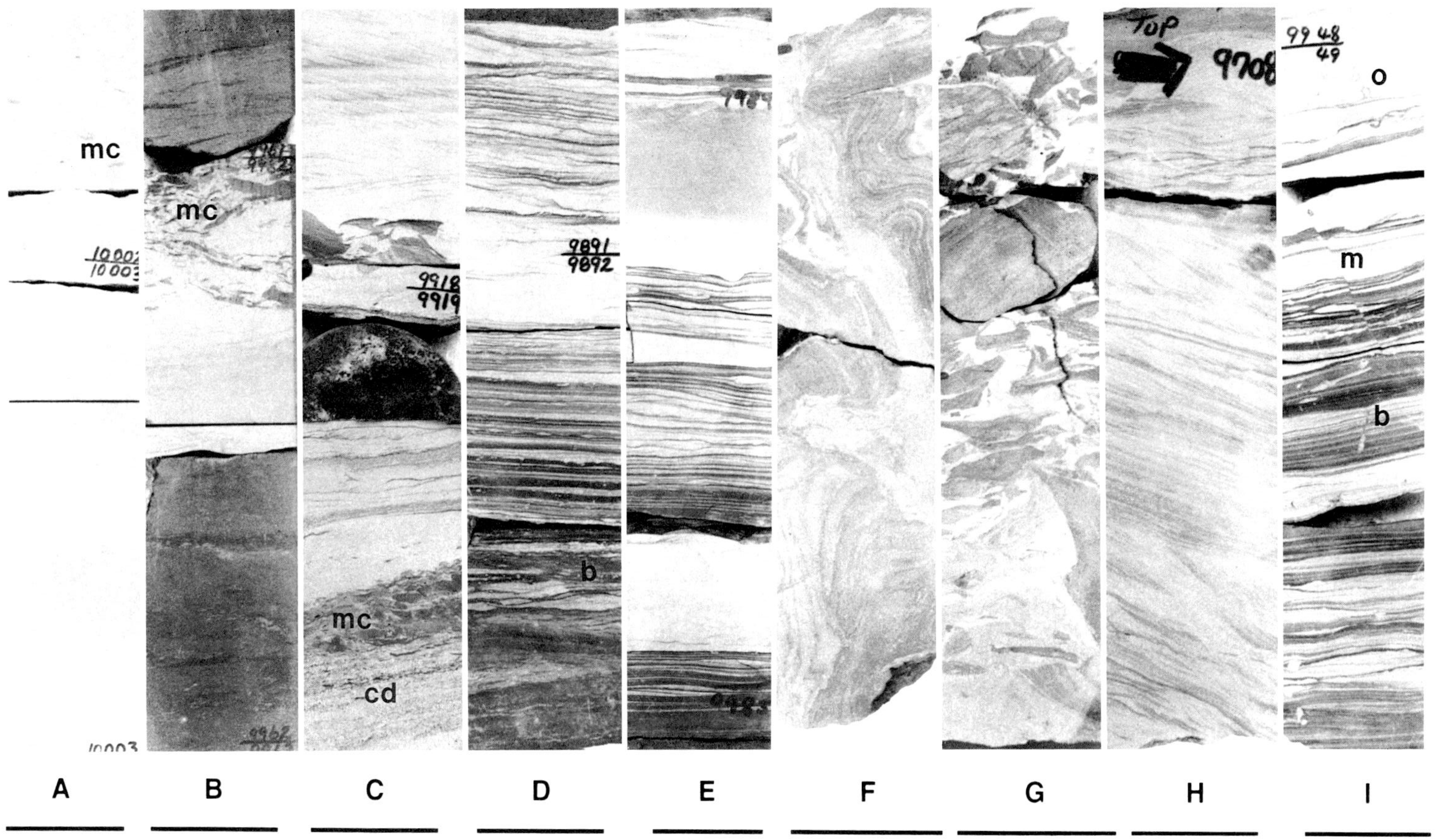

FIG. 16.—Selected features displayed by cores of the Lavaca canyon fill deposits. For each, the well abbreviation (see Fig. 6), lithofacies code, and process facies code are noted. A. HG4 (10002). Sm: CTu. Masssive sandstone with scattered small mud chips (mc). B. HG4 (9981). Sm: LfF. Mud clasts (mc) float within a composite structureless to faintly laminated sandstone bed. Surrounding muddy sandstone shows a diffuse, slurried texture. C. HG4 (9918). Sr: FTr. Abundant carbonaceous debris (cd) and mud clasts (mc) concentrated in ripple-laminated sandstone. D. HG4 (9891). S/Fr, S/Fl: FTu. Microturbidites, common loads, starved and washed-out ripples, and sparse chondrites burrows (b) in thinly interbedded sandstone and mudstone. E. HS1 (9989). S/Fr: FTu/FTr. Low-relief ripple lamination dominates sharp-based, loaded sand beds. Abundance of ripple lamination suggests intermittent traction current reworking of turbidite beds. F. SP1 (9985 ft). Sc: CoF. Contorted and injected muddy sandstone records highly cohesive, viscous flow. G. SP1 (9954 ft). Scl: LfF/CoF. Rounded to angular rip-up clasts in muddy sandstone matrix deposited by cohesive to highly viscous liquified flow. H. HA2. Sr: FTr. Ripple cross-laminated sandstone with multiple cross-cutting scours. I. HA2. S/Fl, S/Fr: FTr. Wavy to ripple-laminated interbedded sandstone and mudstone. Microfaults (m) and burrows (b), including mud-walled *Ophiomorpha* (o) occur at 9949 ft.

incision began. The block records the large-scale mass wasting that excavated the canyon. The displaced sediment display facies typical of the shelf-edge delta front and prodelta environments. Prodelta deposition includes the interplay of mud flows, fine-grained density underflows that formed sandy frontal splays, and hemipelagic settling (Galloway and Hobday, 1983). Delta-front sandstone and sandy laminite are products of traction-current deposition. The massive, structureless, thick sandstone bodies reflect liquefaction that has largely overprinted and obscured primary structures. Spontaneous liquefaction likely accompanied delta-front failure and sliding. The sharp contact and coarse lag at the base of the uppermost sandstone unit suggest that it is a distributary channel fill. A coarse, bioturbated transgressive lag caps the sandstone body at 10,001 ft, again indicating original deposition in shallow water. Overprinting the entire assemblage of deltaic facies is the pervasive 10° to 15° dip and microfaulting. The rotation of a facies succession that is more than 130 ft (40 m) thick and includes part or all of at least three complete delta-front parasequences (the Kubena, Technik, and upper Svetlik sands?) indicates that the core intersects a large slide block. This block rests on the canyon cut surface that outlines the earliest of the canyons of the Lavaca system, the Smothers canyon (Chuber et al., 1989). Rotation left the block tilted into the slide plane. Internal stresses during sliding caused the pervasive microfaulting seen in the core of the deltaic sediment. The Allen #2 slide block reflects a remnant, which remained perched on the upper slope, of the retrogressive shelf-margin collapse that formed the first of the submarine canyons.

Comments: Large slide blocks are a feature of retrogressive basin margins. Analogous early-post-depositional slide blocks of deltaic sediment are described by Nemec et al. (1988) in the Cretaceous of Spitsbergen and by Cossey and Jacobs (1992) in the Oligocene Hackberry Embayment. Where sufficiently large and sandy and entombed in muddy canyon fill, such slide blocks might contain reservoirs. Seismic lines across the Lavaca and nearby Yoakum canyon display rotated slump and slide morphologies within the updip and marginal canyon fill (Galloway et al., 1991).

CONCLUSION

Cores of the Lavaca canyon fill illustrate the array of facies that comprise the relatively muddy Cenozoic slope systems of the northern Gulf of Mexico basin. This facies suite is distinctive only in its setting within a well-defined submarine canyon. The complex facies architecture, lateral variability, and abrupt facies boundaries reflect an origin dominated by gravity mass transport in a submarine canyon, which is inherently a focus of high relief and instability. Channel-fill turbidites are the principal reservoirs within the Lavaca canyon setting. However, given adequate sandstone volume, slide blocks of delta-front deposits and contourite drifts offer reservoir targets in this and other slope depositional sytems. Presence of turbidite lobes basinward remains speculative; if present, they likely lie beyond present drill depth in the deep onlap canyon fill. As in most slope systems, hydrocarbon trapping is inherently stratigraphic and difficult to predict. However, the scale of the channel/lobe complexes and of the slide and slump units is large, making them potentially resolvable on high-resolution 3-D surveys.

ACKNOWLEDGMENTS

Availability of the cores and our understanding of the Lavaca canyon and its geology have benefitted greatly from the generosity of and discussions with Stewart Chuber and Harvey H. Howell. Funds for core retrieval and transport were provided by the Morgan Davis Centennial Professorship, University of Texas at Austin. Drafting was in part by Patrice Porter.

REFERENCES CITED

Berg, R.R., 1979, Characteristics of lower Wilcox reservoirs, Valentine and South Hallettsville fields, Lavaca County, Texas: Gulf Coast Association of Geological Societies Transactions, v. 29, p. 11-23.

Carlson, P.R. and H.A. Karl, 1988, Development of large submarine canyons in the Bering Sea, indicated by morphologic, seismic, and sedimentologic characteristics: Geological Society of America Bulletin, v. 100, p. 1594-1615.

Chuber, S., 1979, Exploration methods of discovery and development of lower Wilcox reservoirs in Valentine and Menking fields, Lavaca county, Texas: Gulf Coast Association of Geological Societies Transactions, v. 29, p. 42-51.

Chuber, S., and R.L. Begeman, 1982, Productive lower Wilcox stratigraphic traps from an entrenched valley in Kinkler field, Lavaca County, Texas: Gulf Coast Association of Geological Societies Transactions, v. 32, p. 255-262.

Chuber, S., H.H. Howell, W.E. Galloway, R.E. Paige, and T.H. Fett, 1989, Lower Wilcox core workshop: Hallettsville Field, Lavaca County, Texas: South Texas Geological Society, unpaginated.

Cossey, S.P.J. and R.E. Jacobs, 1992, Oligocene Hackberry Formation of southwest Louisiana: sequence stratigraphy, sedimentology, and hydrocarbon potential: American Association of Petroleum Geologists Bulletin, v. 76, p. 589-606.

Farre, J.A., B.A. McGregor, W.B.F. Ryan, and J.M. Robb, 1983, Breaching the shelf-break: passage from youthful to mature phase in submarine canyon evolution, *in* D.J. Stanley and G.T. Moore, eds., The shelf break: critical interface on continental margins: Society of Economic Paleontologists and Mineralogists Special Publication 33, p. 25-39.

Galloway, W. E., and D.K. Hobday, 1983, Terrigenous clastic depositional systems: New York, Springer-Verlag, 423 p.

Galloway, W. E., and D.K. Hobday, in press, Terrigenous clastic depositional systems, 2nd Edition: New York, Springer-Verlag.

Galloway, W.E., W.F. Dingus, and R.E. Paige, 1991, Sesimic and depositional facies of Paleocene-Eocene Wilcox Group submarine canyon fills, northwest Gulf Coast, U.S.A., *in* P. Weimer and M.H. Link, eds., Seismic facies and sedimentary processes of submarine fans and turbidite systems: New York, Springer-Verlag, p. 247-271.

Miall, A.D., 1978, Lithofacies types and vertical profile models in braided river deposits: a summary, *in* A.D. Miall, ed., Fluvial sedimentology: Canadian Society of Petroleum Geologists Memoir No. 5, p. 597-604.

Mutti, E., 1992, Turbidite sandstones: Agip S.P.A., 275 p.

Nemec, W., R.J. Steel, J. Gjelberg, J.D. Collinson, E. Prestholm, and I.E. Oxnevad, 1988, Anatomy of collapsed and re-established delta front in lower Cretaceous of eastern Spitsbergen: gravitational sliding and sedimentation processes: American Association of Petroleum Geologists Bulletin v. 72, p. 454-476.

Paige, R.E., 1988, The morphology, depositional setting, and evolution of the Paleocene Lavaca submarine canyon system, northwest Gulf of Mexico: M.A. Thesis, The University of Texas at Austin, 123 p.

Pickering, K.T., R.N. Hiscott, and F.J. Hein, 1989, Deep marine environments: London, Unwin Hyman, 416 p.

Reading, H.G., 1991, The classification of deep-sea depositional systems by sediment calibre and feeder system: Journal of the Geological Society of London, v. 148, p. 427-430.

Reading, H.G., and Marcus Richards, 1994, Turbidite systems in deep-water basin margins classified by grain size and feeder system: American Association of Petroleum Geologists Bulletin v. 78, p. 792-822.

Shanmugam, G., T.D. Spalding, and D.H. Rofheart, 1993, Process sedimentology and reservoir quality of deep marine bottom-current reworked sands (sandy contourites): an example from the Gulf of Mexico: American Association of Petroleum Geologists Bulletin v. 77, p.1241-1259.

Stow, D.A.V., 1986, Deep clastic seas; *in* H.G. Reading, ed., Sedimentary environments and facies: Oxford, Blackwell Scientific Publications, p. 399-444.

Van Wagoner, J.C., R.M. Mitchum, K.M. Campion, and V.D. Rahmanian, 1990, Siliciclastic sequence stratigraphy in well logs, cores, and outcrops: concepts for high-resolution correlation of time and facies: American Association of Petroleum Geologists Methods in Exploration Series, No. 7, 55 p.

Xue, L., and W.E. Galloway, 1993, Sequence stratigraphic and depositional framework of the Paleocene lower Wilcox strata, northwest Gulf of Mexico Basin: Gulf Coast Association of Geological Societies Transactions, v. 43, p. 453-463.

DEEP-MARINE BOTTOM-CURRENT REWORKED SAND (PLIOCENE AND PLEISTOCENE), EWING BANK 826 FIELD, GULF OF MEXICO

G. SHANMUGAM[1], T. D. SPALDING[2], and D. H. ROFHEART[2]
[1]*Mobil Research and Development Corporation, P.O. Box 650232, Dallas, Texas 75265-0232*
[2]*New Business Development, P.O. Box 650232, Dallas, Texas 75265-0232*

ABSTRACT

The Pliocene-Pleistocene sequence cored in the Ewing Bank 826 Field in the Gulf of Mexico provides an example of sand distribution and reservoir quality produced by reworking by deep-marine bottom currents. A distinctive attribute of reworked sands is their traction bedforms. Common sedimentary features of traction currents include small-scale cross-bedding, starved current ripples, horizontal lamination, sharp upper contacts, and inverse size grading. The sands also exhibit internal erosional surfaces and mud-offshoots indicating oscillating current conditions.

Presumably, the Loop current, a strong wind-driven surface current in the Gulf of Mexico, impinged on the sea bottom, as it does today, and reworked sand. A depositional model based on the integration of core, wireline log, and 3-D seismic data suggests that the reworked sediment package may be thick and continuous, but individual sand layers within the package may be thin and discontinuous. This model, which depicts the distribution of bottom-current reworked sand in interchannel slope areas as a distinctly different facies from channel-levee facies, has the potential for general application to other deep-water plays outside the study area. In the Ewing Bank 826 Field, the Type 1 (L-1) reservoir with 80% sand exhibits higher permeability values (100-1800 mD) than the Type 2 (N-1) reservoir with 26% sand (50-800 mD). The increased permeability in the Type 1 sand has been attributed to high sand content, vigorous reworking, and microfractures. The clean, porous and well-sorted Type 1 sands with good communication between sand layers have produced at higher rates and recovery efficiencies than the Type 2 sands with numerous interbedded mud layers.

INTRODUCTION

Recent studies have recognized the influence of bottom-currents in the hydrocarbon-bearing Lower Tertiary sands of the Frigg Field, North Sea (Enjolras et al., 1986), in the Eocene sands of the Campos Basin of the Brazilian offshore (Mutti et al., 1980), and in Pliocene and Pleistocene intraslope reservoirs of the northern Gulf of Mexico (Shanmugam et al., 1993a, 1993b). Because sand deposited by bottom currents can form hydrocarbon-bearing reservoirs, a clear understanding of their depositional mechanisms, facies characteristics, sand-body geometry, and reservoir quality is essential for effective exploration and exploitation.

In spite of continuing interest in bottom currents and contourites (Mutti, 1992; Stow and Faugeres, 1993), distinguishing bottom-current reworked sands from turbidites and other gravity generated deep-marine sands is still an unresolved issue in deep-marine sedimentology. The objectives of this study are: (1) to document primary sedimentary features in core from the Ewing Bank 826 Field and the adjacent Garden Banks areas indicative of bottom-current reworking; (2) to develop a depositional model for bottom-current reworked Pliocene-Pleistocene sand based on the integration of conventional core, wireline log, and 3-D seismic data in the Ewing Bank 826 Field (Fig. 1); (3) to discuss the importance of traction transport associated with bottom currents; and (4) to discuss reservoir geometry and quality of reworked sand.

BOTTOM-CURRENT REWORKED SAND

Bottom Currents

The presence of deep bottom currents in the modern deep sea has been documented by numerous studies from current measurements, bottom photography, piston coring, and high-resolution seismic profiling. In large modern ocean basins, such as the Atlantic, deep, thermohaline-induced geostrophic bottom currents (Ewing et al., 1971; Flood and Hollister, 1974) flow parallel to bathymetric contours (i.e., along slope, Fig. 2) and are generally referred to as "contour currents" (Heezen et al., 1966). However, because not all types of bottom currents follow regional bathymetric contours, we prefer the term "bottom current" to "contour current." For example, wind-driven surface currents may flow in a circular motion and may reach the deep-sea floor, such as the Loop Current in the Gulf of Mexico (Pequegnat, 1972; Bouma, 1972) and the Gulf Stream in the North Atlantic. In addition, local bottom currents can be associated with by tides and internal waves; especially in submarine canyons (Bouma and Hollister, 1973; Shepard et al., 1979).

The speed of bottom currents can be several knots (1 knot = 51 cm/sec). Measured velocities usually ranges from 1-20 cm/sec (Hollister and Heezen, 1972); however, exceptionally strong near-bottom currents with maximum velocities of up to 300 cm/sec have been recorded in the Straits of Gibraltar (Gonthier et al., 1984). Bottom-current velocities of 73 cm/sec have been measured in 5000 m water depths on the lower continental rise off Nova Scotia (Richardson et al., 1981). These currents are quite capable of erosion, transportation, and redeposition of fine- to coarse-grained sand in the deep sea. Regional erosional unconformities throughout thousands of square miles of sea floor can also be created by bottom currents in the deep

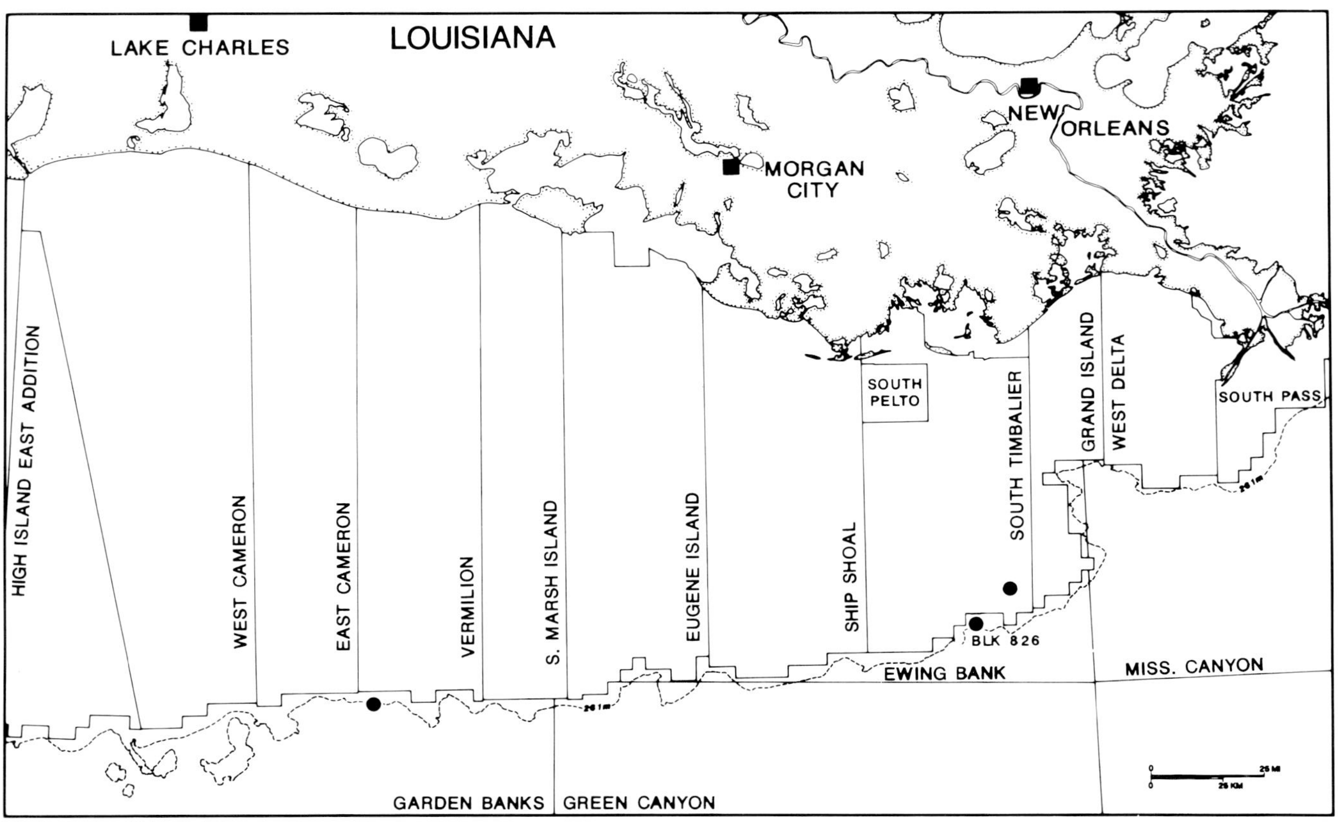

FIG. 1.—Location map of the Ewing Bank and adjacent areas in the northern Gulf of Mexico. Dots show approximate location of cores used in this study.

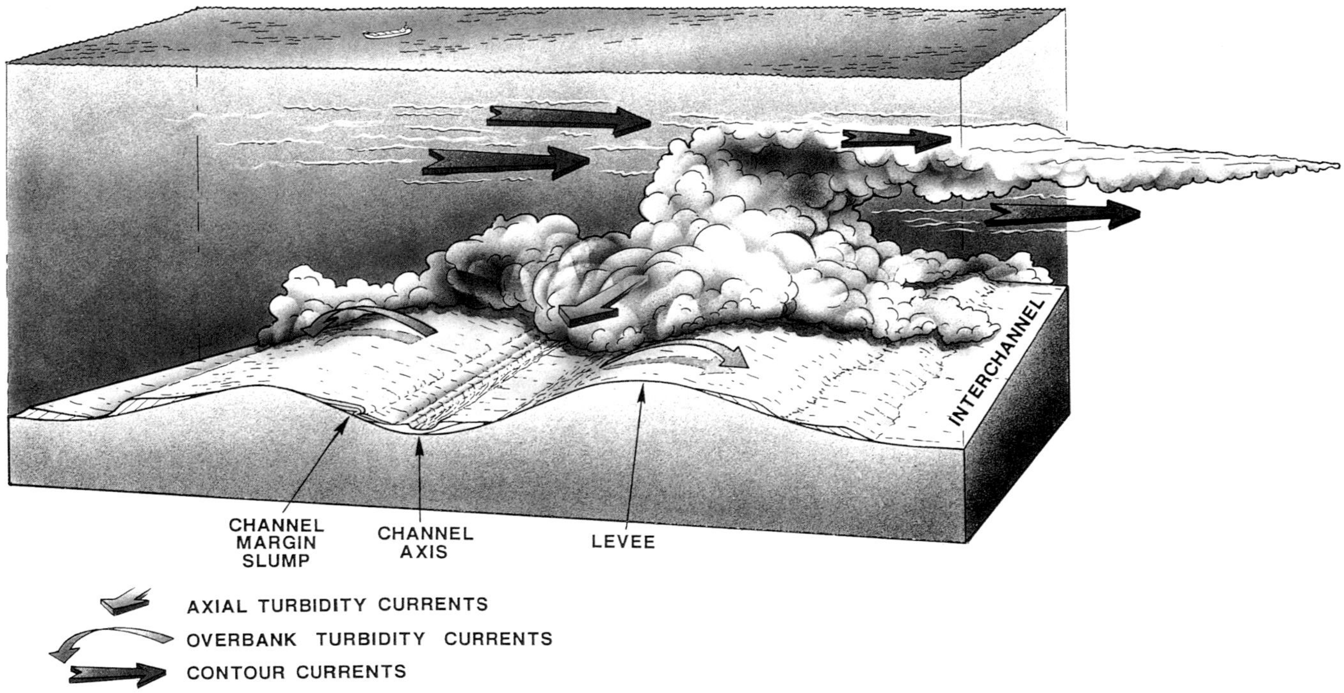

FIG. 2.—A conceptual model showing spatial relationship between axial turbidity currents, overbank turbidity currents, and bottom currents (contour currents) in deep-sea environments. Note downslope axial turbidity currents and along slope bottom currents (contour currents). Bottom-current reworking of levee sands may be common in the interchannel areas.

sea (Berggren and Hollister, 1977; Tucholke and Embley, 1984).

Bottom currents: (1) generally are persistent for long time intervals when compared to turbidity currents; (2) sometimes are free of sediment, and for this reason, are termed "clear-water currents" (Bouma and Hollister, 1973, p. 82); (3) are driven by thermohaline, wind, wave, or tidal forces; (4) entrain and transport passive sediment particles (Lowe, 1979); (5) commonly flow parallel to the regional slope, but can also flow in circular gyres or oblique to bathymetry (Pequegnat, 1972); and (6) transport sand primarily by tractive bedload movement as sliding, rolling, and salting grains (Allen, 1982). These characteristics differentiate deep-sea bottom currents from turbidity currents.

Bottom-Current Deposits

Deposits of "contour currents" (i. e., bottom currents) have been termed "contourites" (Hollister and Heezen, 1972). Two types of contourites, namely "muddy contourites" and "sandy contourites", have been recognized (Stow and Lovell, 1979; Stow and Holbrook, 1984). The muddy type settles directly from sluggish nepheloid bottom currents, and the sandy type represents sands, and even gravels, that have been reworked and winnowed of fines by more vigorous bottom currents. Although criteria have been proposed to distinguish muddy from sandy contourites (Stow and Lovell, 1979), the real distinction between the two types in terms of sand content has not been established. There exists a continuum in grain size between muddy and sandy types. We prefer the term "bottom-current-reworked sands" to "sandy contourite" irrespective of the sand content, so long as evidence for reworking of sand can be established. All core examples described in this study may be classed as "sandy contourites" in the classification of Lovell and Stow (1981).

Bottom-current-reworked sands can be recognized using primary physical sedimentary structures. Criteria for recognizing bottom-current deposits were presented by Hubert (1964), Hollister (1967), Hollister and Heezen (1972), Bouma and Hollister (1973), Unrug (1977), Stow and Lovell (1979), Lovell and Stow (1981), Shanmugam (1990), Shanmugam et al. (1993a, 1993b). The importance of some of the criteria suggested by earlier workers has been minimized as more cases of reworked sands are recognized. For example, bioturbation was considered a diagnostic criterion to recognize sandy contourites (Lovell and Stow, 1981). This was based on the belief that active bottom currents would increase the oxygen concentration of the water mass (Chough and Hesse, 1985), and thereby would increase the activity of organisms. However, Tucholke et al. (1985) argued that the amount of bioturbation is in part an index of bottom current intensity; strong bottom currents do not favor preservation of biogenic structures. Therefore, absence of bioturbation in deep-sea sediment cannot be used as evidence against the existence of bottom currents. In the cases we have studied, bioturbation is not important.

The use of paleocurrent directions to recognize bottom-current deposits (Lovell and Stow, 1981) under the assumption that all bottom currents follow regional bathymetric contours may also be tenuous. Surface currents that flow in circular motions (gyres) unrelated to the slope contours, such as the Loop Current in the Gulf of Mexico (Pequegnat, 1972), can often impinge on the sea floor. Tidal currents can flow up and down the submarine canyons, perpendicular to the slope (Shepard et al., 1979). Current reversals have also been associated with abyssal storms (Hollister and McCave, 1984).

We suggest that primary physical sedimentary structures are better indicators of reworked sands than bioturbation and paleocurrent directions. A synthesis of recent work suggests the following general criteria for recognizing bottom-current-reworked sand in core:

1. Predominantly fine-grained sand and silt.
2. Thin-bedded to laminated sand (usually less than 5 cm thick) (Fig. 3).
3. Rhythmic occurrence of sand and mud layers (Fig. 3).
4. Numerous layers (50 or more per 1 meter of core) (Fig. 3).
5. Sharp (non-erosional) upper contacts (Fig. 3) and sharp to gradational bottom contacts.
6. Internal erosional surfaces (Figs. 4 and 5).
7. Well-sorted sand and low depositional matrix.
8. Inverse size grading within beds (coarsening upward) at various scales (Figs. 6 and 7).
9. Horizontal lamination and low-angle cross lamination (Fig. 3).
10. Cross-bedding (Figs. 3, 4, 5, 8).
*11. Cross beds accentuated by heavy minerals.
12. Lenticular bedding/starved ripples at the core scale (Fig. 3).
13. Current ripples with preserved crest (Fig. 9) or eroded crest (Fig. 9).
14. Ripple forms with curved bases or wave ripples (Fig. 9).
15. Mud-offshoots (Figs. 10, 11).
16. Flaser bedding (Fig. 12).

*Not recognized in the Ewing Bank cores

We emphasize that no single criterion by itself is unique to bottom-current reworked facies. Although many of the features listed above can be attributed to processes other than bottom-current reworking, the association of several of the above features such as observed in the Ewing Bank cores, greatly enhances the chance for recognizing bottom-current reworked facies.

Importance of Traction

Most of the diagnostic features listed above are products of traction or combined traction and suspension deposition (Fig. 13). Traction features are an indication of current reworking of granular grains by bottom currents. The fact that the traction features occur by themselves without forming part of a regular vertical sequence of structures, such as the graded Bouma sequence, suggests processes other than waning turbidity currents.

Cross-bedding is common in the cores examined (Figs. 3, 4, 5, 8). Ripple wave lengths of cross laminae range from 10 to 20 cm and amplitudes are less than 2 cm. The foreset thickness is in the range of 4 or 5 cm, which suggests long, stable periods of bedload movement. Cross-bedding is the strongest evidence for bottom-current-reworking. This is because avalanching and current traction are more likely to occur under "clear water" bottom currents than under sediment-charged turbidity currents. Also, sustained flow conditions, required for the formation of cross beds, are more common in bottom currents than in turbidity currents. More importantly, these cross-strata are not associated with turbidite sequences that show grading or the Bouma sequence.

Although Mutti (1977) advocated the origin of cross-bedding in deep-sea channel-mouth deposits due to by-passing of turbidity currents, quasi-steady flow conditions required for the formation of cross-bedding are difficult to envision in episodic turbidity current events at channel mouths.

Dip of cross-bedding appears to vary from bi-directional (Fig. 8) to multi-directional (Fig. 9). Conceivably,

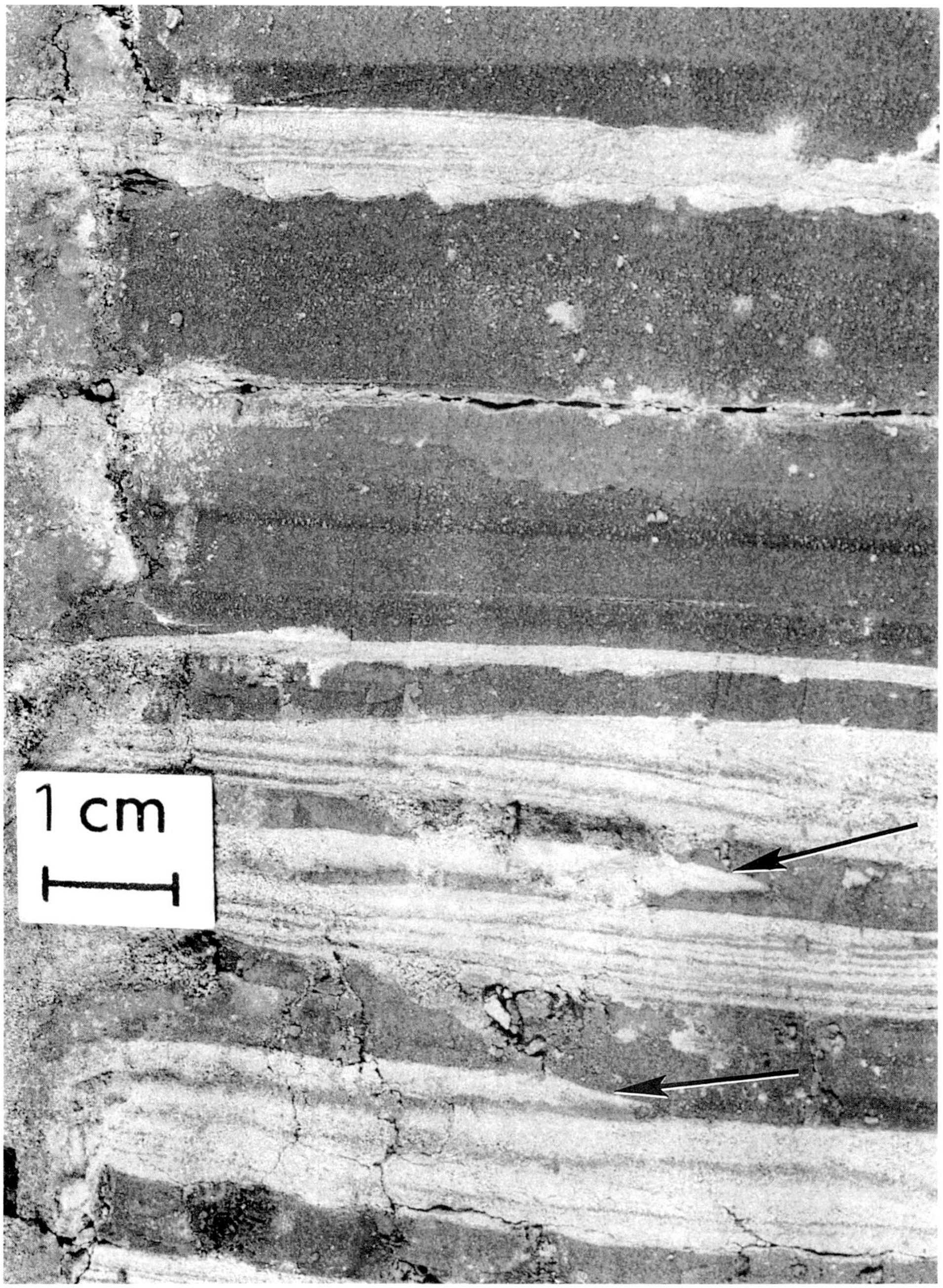

FIG. 3.—Core photograph showing discrete thin sand layers with sharp upper contacts. Traction bedforms include horizontal laminae, starved ripples (arrows), and low-angle cross-bedding (sand unit above the upper arrow). Dip of cross- bedding to the right suggests current flow was from left to right in this example. Note rhythmic occurrence of sand and mud layers. Middle Pleistocene, Gulf of Mexico.

FIG. 4.—Core photograph showing cross-bedded units with faint internal truncation surfaces (arrows). Beds dip to the left. Middle Pleistocene, Ewing Bank Block 826, Gulf of Mexico.

FIG. 5.—Core photograph showing an erosional truncation surface (arrow) between cross-laminated sand unit and overlying mud unit. Upper Pliocene, Ewing Bank Block 826, Gulf of Mexico.

tide-related bottom currents that flow up and down submarine canyons (Shepard et al., 1979) may have caused this bi-directional cross bedding. An Ordovician example in Virginia with a bi-directional cross-laminated unit in a submarine channel environment has been attributed to internal tide origin (Zhenzhong and Eriksson, 1991).

Another cause of bi-directional cross bedding is reworking of levee deposits by bottom currents. For example, if overbank turbidity currents form ripple laminae on both banks of a channel (Fig. 2), bottom currents that flow in a direction orthogonal (90°) to those turbidity currents on the left-hand bank would rework levee sand. Such cross beds can also be explained by the reversal of current direction resulting from eddy circulation associated with abyssal storms (Hollister and McCave, 1984).

The erosional truncation surfaces or reactivation surfaces, which occur within cross-laminated units, reflect pulses of increased energy within bottom currents. Such alternating episodes of traction deposition and erosion are not typical of a turbidity current regime. Oscillating energy conditions of bottom currents have been observed and measured over the HEBBLE (High-Energy Benthic Boundary Layer Experiment) site on the Nova Scotia continental rise (Hollister and McCave, 1984). Strong fluctuations of bottom currents can create local erosion (Flood and Shor, 1988).

The sharp upper contacts of sand layers (Fig. 3) have been attributed to reworking by bottom-currents (Bouma and Hollister, 1973) because of the oscillating energy conditions (Gonthier et al., 1984). There is no evidence to suggest that the sharp upper contacts are caused by erosion (Fig. 3). The sharp upper contacts are not typical of turbidites because deposition from suspension in turbidity currents results in gradational upper contacts. We suggest

FIG. 6.—Core photograph showing larger scale inverse size grading. Arrow (12 cm long) shows gradual upward increase in grain size from mud (dark) at bottom to sand (light) at top. Note gradational lower contact of sand unit. Middle Pleistocene, Gulf of Mexico.

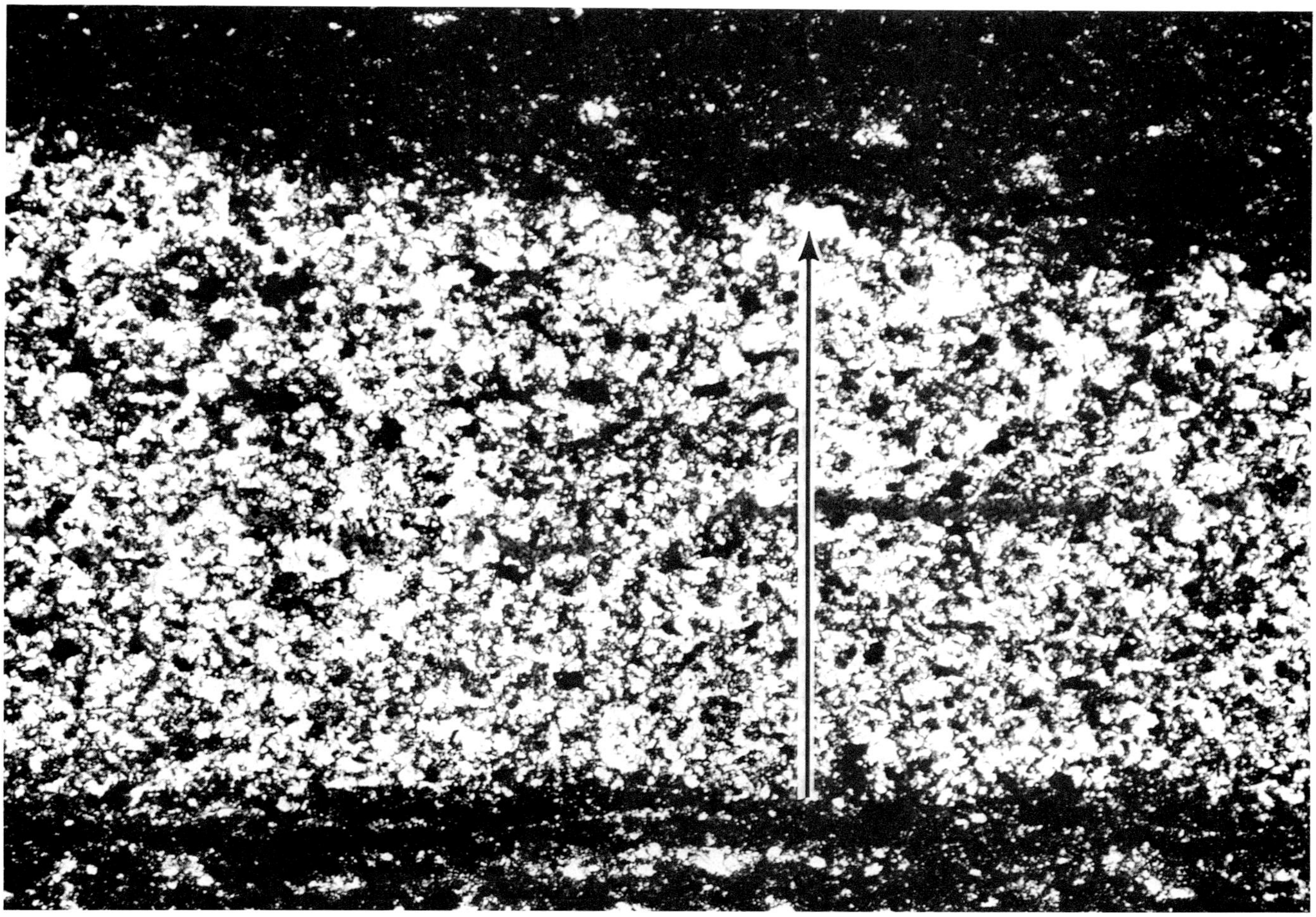

FIG. 7.—Photomicrograph of a fine-grained sand layer (arrow) showing smaller scale inverse size grading and sharp upper contact. Note largest quartz grain (tip of arrow) at the top of this 1 mm thick sand layer. Middle Pleistocene, Gulf of Mexico.

that contemporaneous bottom currents (contour currents) continually reworked and redistributed much of the sediments introduced by turbidity currents or other sediment gravity flows into the thin sand layers with sharp upper contacts observed in these cores. The time interval between successive turbidity current events was sufficient to permit complete reworking of the thin turbidite sand layers. In cases of thick turbidite sand layers, however, reworking may not be complete.

Lenticular bedding with starved ripple forms (Fig. 3) is common in bottom-current deposits. Such beds are usually less than 1 cm thick. Lenticular laminae of bottom-current origin have been reported from DSDP leg 28, Site 268, in Antarctica (Piper and Brisco, 1975). Starved ripples are more likely to be formed by relatively sediment free "clear-water" bottom currents than by sediment-laden turbidity currents.

Ripple cross-laminae with preserved mud streaks in their troughs, popularly known as flaser bedding (Reineck and Singh, 1980), have been observed in cores (Fig. 12). Thin (1-2 mm thick) mud layers occasionally drape over the entire sand ripple, which sometimes resemble a "sigmoidal" shape (Figs. 10 and 11). We have termed these mud drapes on ripples or ripple foresets as "mud-offshoots." Current ripples with mud drapes have been observed on the modern ocean floor in the Gulf of Mexico (Pequegnat, 1972). The formation of mud-offshoots in a rippled sand may be explained by alternating high and low energy conditions. During periods of high energy, current activity maintains ripples (i. e., avalanching phase), whereas

FIG. 8.—Core photograph showing bi-directional cross-laminae (highlighted by two dashed lines) in sand. Upper Pliocene, Ewing Bank Block 826, Gulf of Mexico.

during periods of low energy, mud deposited from suspension drapes over the ripple foresets. The mud-offshoots represent low-angle foresets caused by settling of mud from suspension (i. e., non-avalanching phase) on the lee side of ripples. Mud drapes are commonly reported from nearshore and offshore tidal environments, but they have not been reported from deep-sea turbidites. This is because fluid turbulence in turbidity currents would prevent formation of the mud drapes.

Some current ripples show eroded crests (Figs. 9 and 10). These truncation surfaces represent removal of sediment from the ripple crest during strong traction. Under strong energy conditions, vigorous bedload movement may lead to a "high velocity plane bed."

Conventionally, wave ripples are considered indicators of shallow-water environments. Ripples with curved bases (Fig. 9) may be considered as wave ripples, but occur as part of a deep-water (upper bathyal) sequence. One possible explanation might be internal waves. For example, Karl et al. (1986) suggested that during periods of low sea level delivery of large volumes of fresh water and large quantities of sediment directly to the slope may enhance the propagation of high-frequency internal waves. Such conditions might explain the presence of wave ripples in deep-water sequences. Although these deep-marine sands resemble tidal and shallow marine (outer shelf) deposits, they commonly occur in close association with other deep-marine facies, such as turbidites, debris-flow deposits, and hemipelagic mud.

Although rippled units in deep-water sequences are routinely interpreted as "levee" facies by many workers, it is important to establish that the rippled units are indeed products of overbank turbidity currents in justifying a levee interpretation. In levee facies, rippled units com-

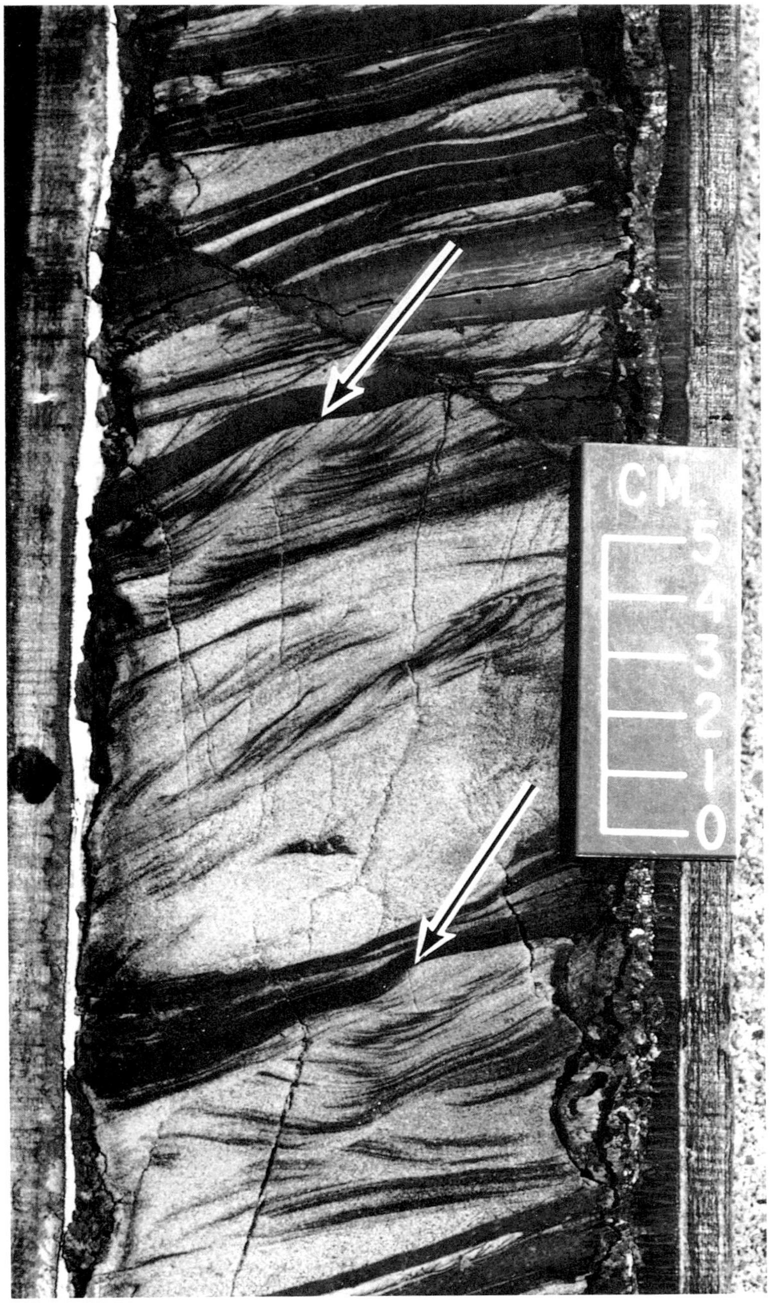

FIG. 9.—Core photograph showing discrete sand units comprised of current ripples with variable dip directions suggesting multiple current directions. Preserved (lower arrow) and eroded (upper arrow) tops of ripples indicate variable energy conditions of the current. Ripples with curved bases probably indicate wave influence. Middle Pleistocene, Ewing Bank Block 826, Gulf of Mexico.

FIG. 10.—Core photograph showing discrete sand units with current ripples and mud-offshoots (dark layers dipping to the left pointed by arrow). Note sigmoidal configuration of ripples and truncated tops. Middle Pleistocene, Ewing Bank Block 826, Gulf of Mexico.

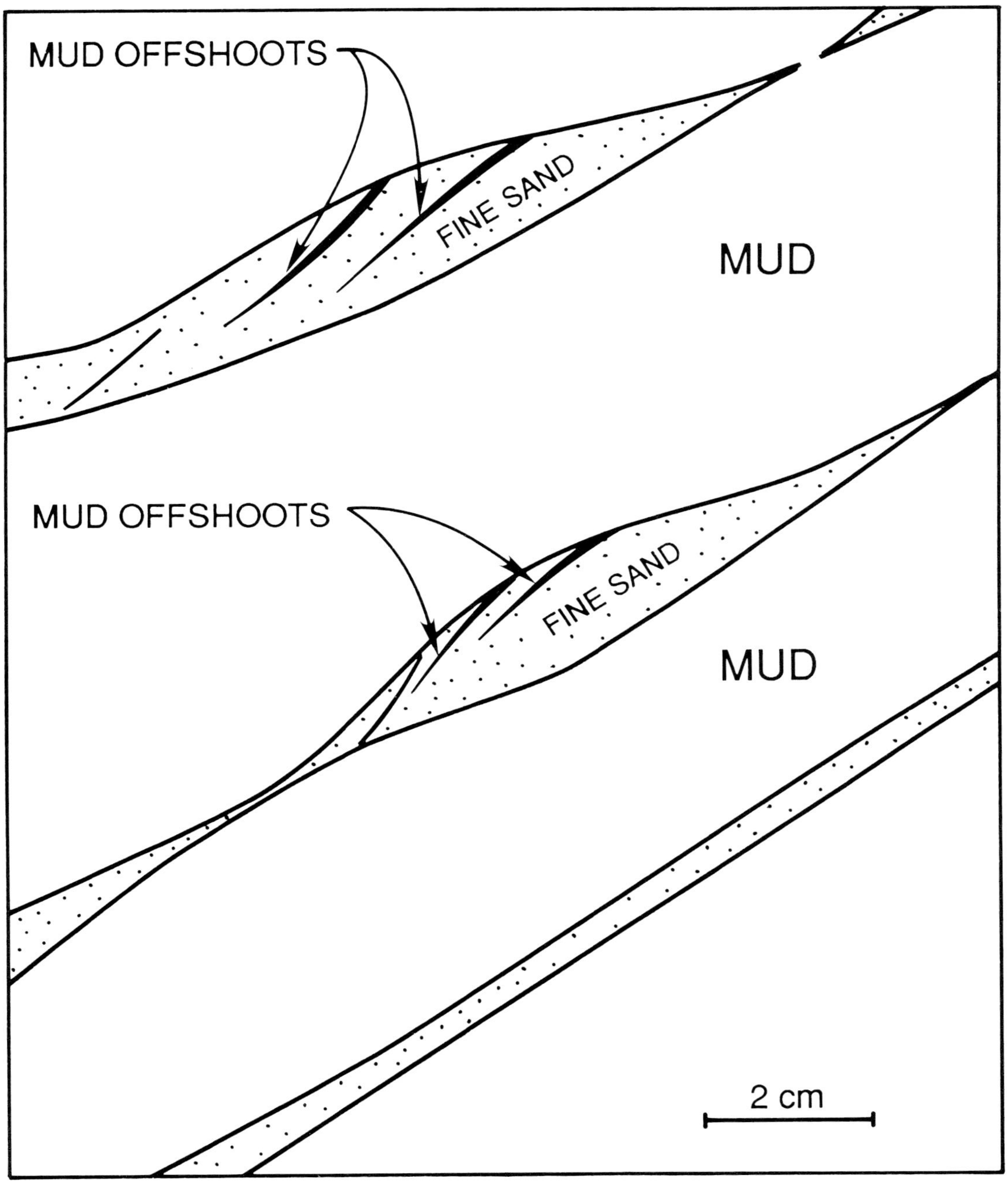

FIG. 11.—Sketches of mud offshoots in fine-grained sand layers.

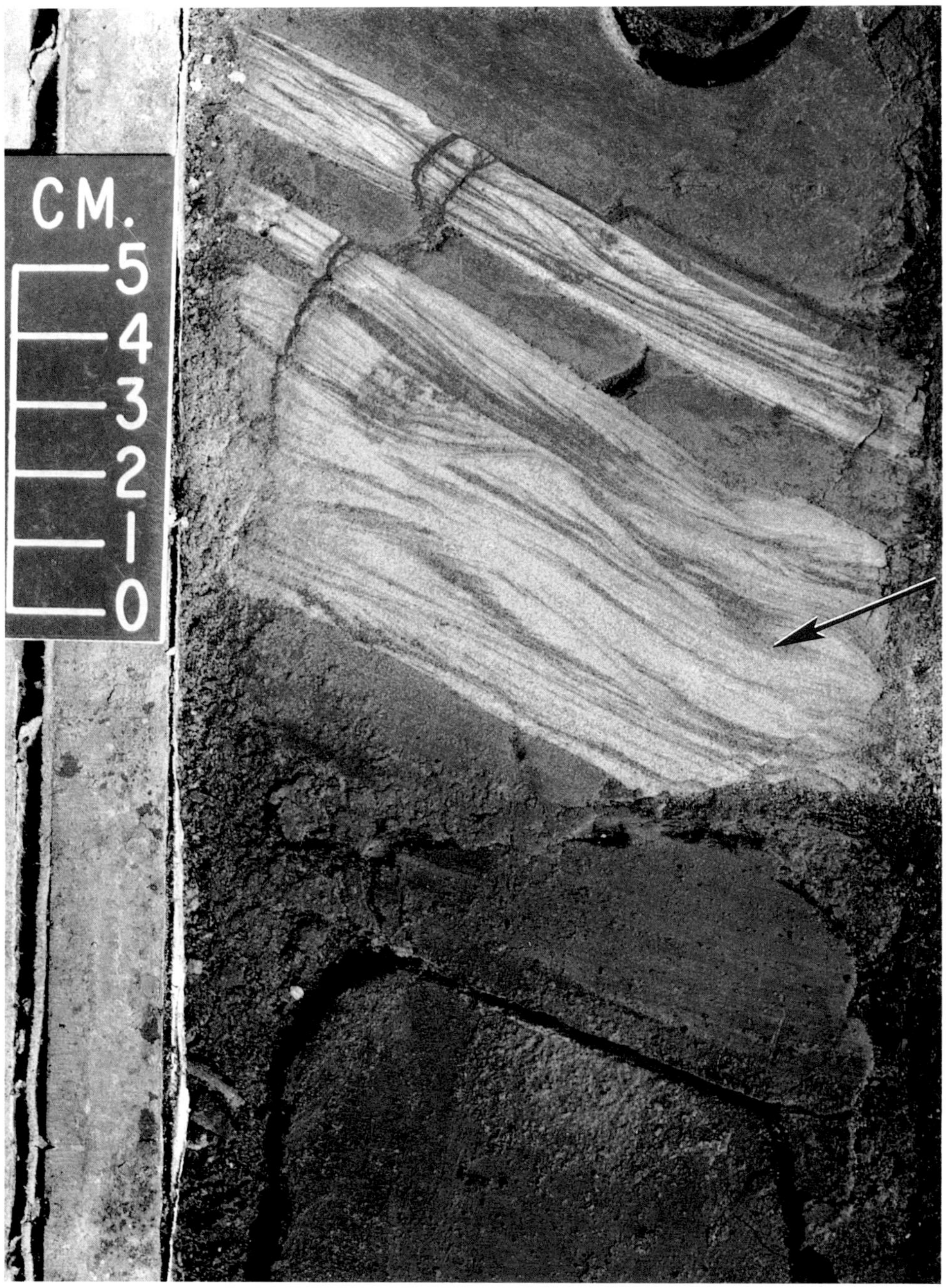

FIG. 12.—Core photograph showing flaser bedding. Note the presence of mud in ripple troughs (arrow). Upper Pliocene, Ewing Bank Block 826, Gulf of Mexico.

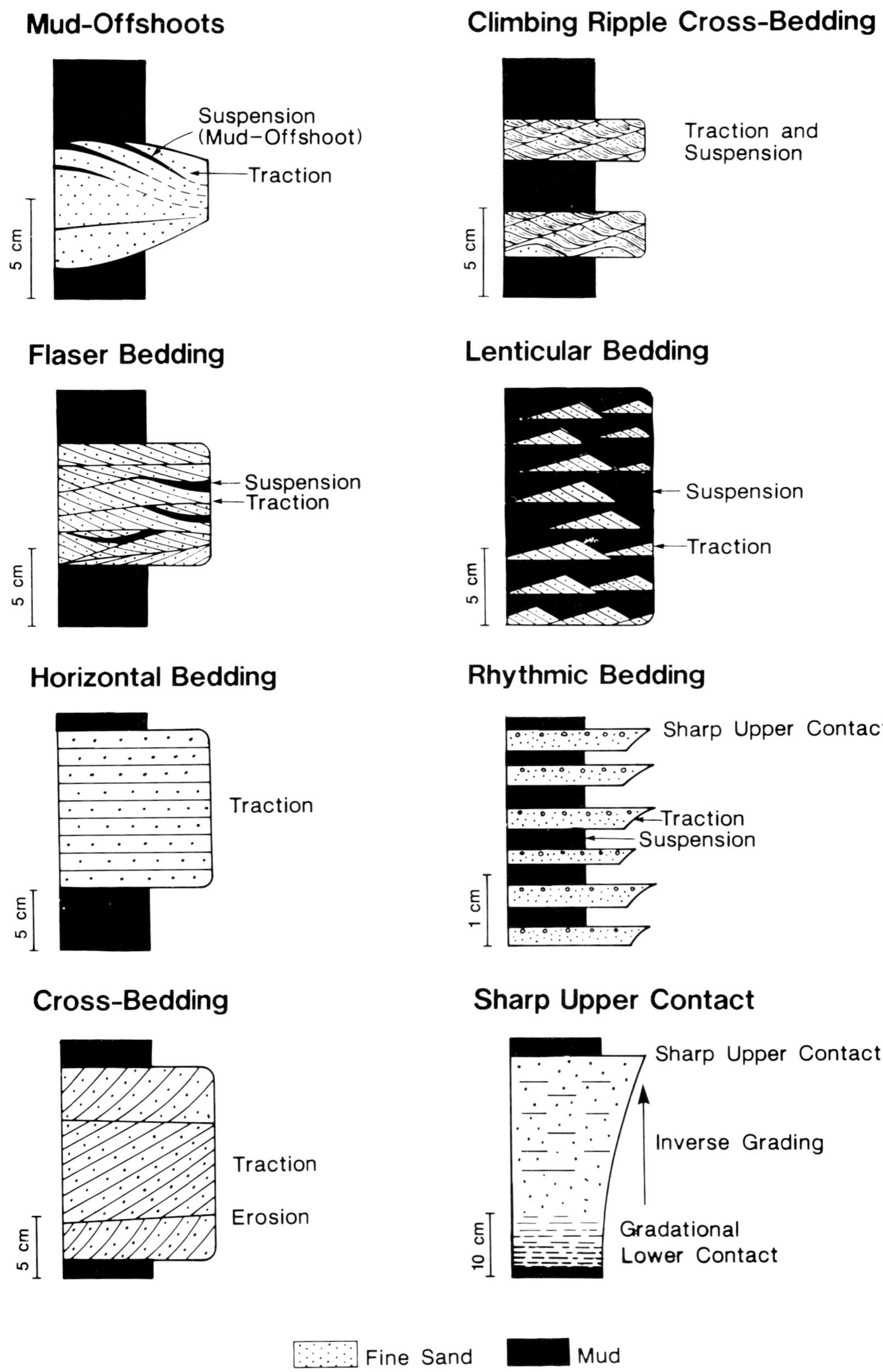

FIG. 13.—Generalized facies models for deep-marine bottom-current reworked sands showing traction bedforms.

monly occur as a part of a graded turbidite sequence. Because rippled units in our study area occur as discrete units, we do not interpret them as products of overbank turbidity currents. Furthermore, cross bedding, horizontal lamination, mud offshoots, and inverse size grading observed in the core are not characteristic of levee facies (Fig. 14). We suggest that bottom-current reworked sands can develop in slope, rise, and basin plain areas with or without channels, whereas levees are typically associated with aggradational channel facies.

EWING BANK 826 FIELD, GULF OF MEXICO

The Ewing Bank Block 826 Field, located nearly 100 km (62 miles) off the Louisiana coast in the northern Gulf of Mexico (Fig. 1), contains hydrocarbon-producing clastic reservoirs that we interpret as being formed by deep bottom currents. The field was discovered in May of 1985, and lies in approximately 150 m (500 feet) of water along the "flexure trend" on the present day outer shelf and upper continental slope. The field is operated by BP Exploration (formerly Sohio), and the partnership includes Mobil (40%), BP Exploration (40%), and Kerr-McGee (20%). A structure map on top of the Lower Pleistocene (LPL) is shown in Figure 15. A schematic cross section across the field (A to A' in Fig. 15) is shown in Figure 16 with the two Upper Pliocene reservoir sands (L-1 and N-1) discussed in this paper. Examples of reworked sand from the Middle Pleistocene (MPL in Fig. 16) sequence are also used in this study.

Six wildcat and appraisal wells were drilled during 1985 and 1986. Four cores (12,594.00-13,328.30 ft) of Upper Pliocene age from Sohio's Ewing Bank 826, OCS-G-5800 Well No. 3, and two cores (13,360.00-13,416.70 ft) from Well No. 3ST (Sidetrack) (Fig. 17) were described for this study. A four inch conventional core barrel was used with a fiberglass liner. The No. 3 well is a straight hole in which log depths equal core depths. The sidetrack well is deviated about 30° from vertical, and log depth is 13 feet deeper than core depth. Cores were examined at Petroleum Testing Service, Inc., Santa Fe Springs, California, where the cores were kept in frozen storage.

The entire cored sequence may be considered to show a "crescent-shaped" or a "Christmas-tree" log motif (Fig. 17), indicating a slope fan composed of channel-levee systems (Vail and Wornardt, 1990; Mitchum et al., 1990). However, sedimentary structures from the L-1 sand of the 3ST well reveal three distinctly different depositional facies, namely, a lower levee unit, a middle turbidite channel sand unit, and an upper bottom-current reworked sand unit (Fig. 17). The levee unit is characterized by slumped mud, the turbidite channel unit by graded sand beds displaying Bouma divisions, and the bottom-current reworked sand unit by current ripples and sharp upper contacts. Thus in the absence of cores, the interpretation of depositional facies using log motifs alone can be misleading and could result in unrealistic depositional models.

Bottom-current reworked sands have also been recognized in the Pliocene and Pleistocene intraslope reservoirs in the Garden Banks, South Timbalier, High Island A-474/A-499 (Anspach et al., 1989), and Green Canyon areas (Fig. 1). Two other evidences suggest that bottom currents could have been operative in the Gulf of Mexico during the Pleistocene and Holocene times: (1) DSDP Leg 96 latest Pleistocene cores on the Mississippi Fan at Site 614A reveal evidence for bottom-current reworking in the form of sand layers with sharp upper contacts, inverse grading, and cross-laminae (Shanmugam et al., 1988). (2) Fields of sediment waves have been observed along the base of the Sigsbee Escarpment (Damuth et al., 1991).

The Loop Current

We infer that in the Ewing Bank area, wind-driven surface currents apparently penetrated to the sea floor during the Pliocene-Pleistocene and were responsible for the current-reworked sands observed in the Ewing Bank and adjacent areas. Deep-thermohaline-driven water masses cannot be responsible because such water masses are unable to enter the Gulf of Mexico due to sea floor topographic barriers in the Caribbean basins. In the eastern Gulf of Mexico, oceanic currents are dominated by the modern Loop Current (Nowlin, 1972). Advanced Very High Resolution Radiometry (AVHRR) satellite image of the Loop Current (Fig. 18) shows that this current enters the Gulf through the Yucatan Straight as the Yucatan Current, flows in a clockwise loop in the eastern Gulf as the Loop Current, and exits via the Florida Strait as the Florida Current (Fig. 19). Finally, this current merges with the Antilles Current to form the Gulf Stream. The Loop Current also propagates eddies into the north-central Gulf of Mexico where the Ewing Bank area is located (Fig. 19).

Cooper et al. (1990) recorded velocities as high as 4 knots (204 cm/sec) at 100 m depths in eddies recently detached from the Loop Current. The Loop Current and its eddies pose significant problems to deep-water drilling (Koch et al., 1991). For example, drilling operations in the Green Canyon 166 area were temporarily suspended in

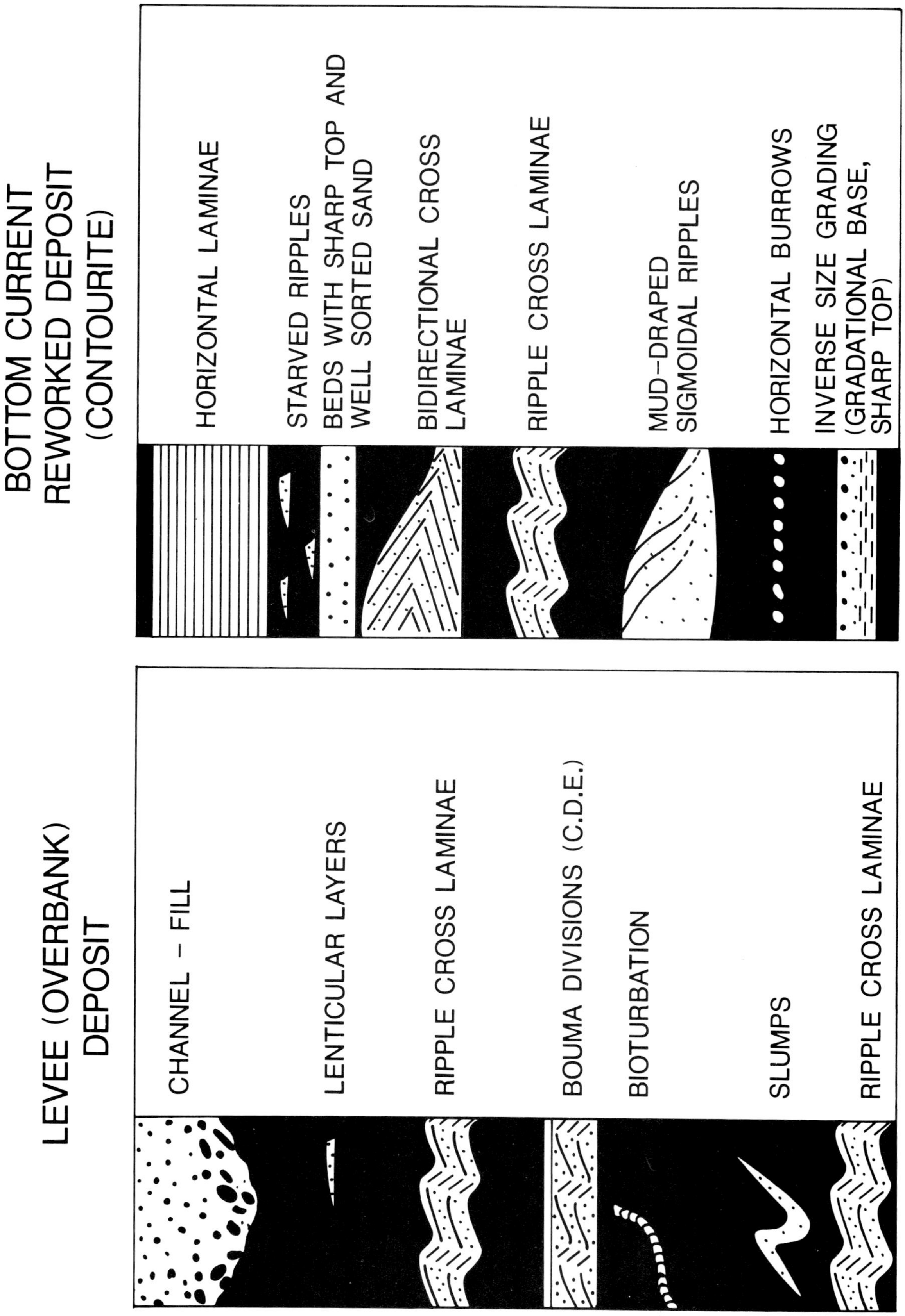

FIG. 14.—Generalized facies models showing differences between levee and bottom-current reworked sands.

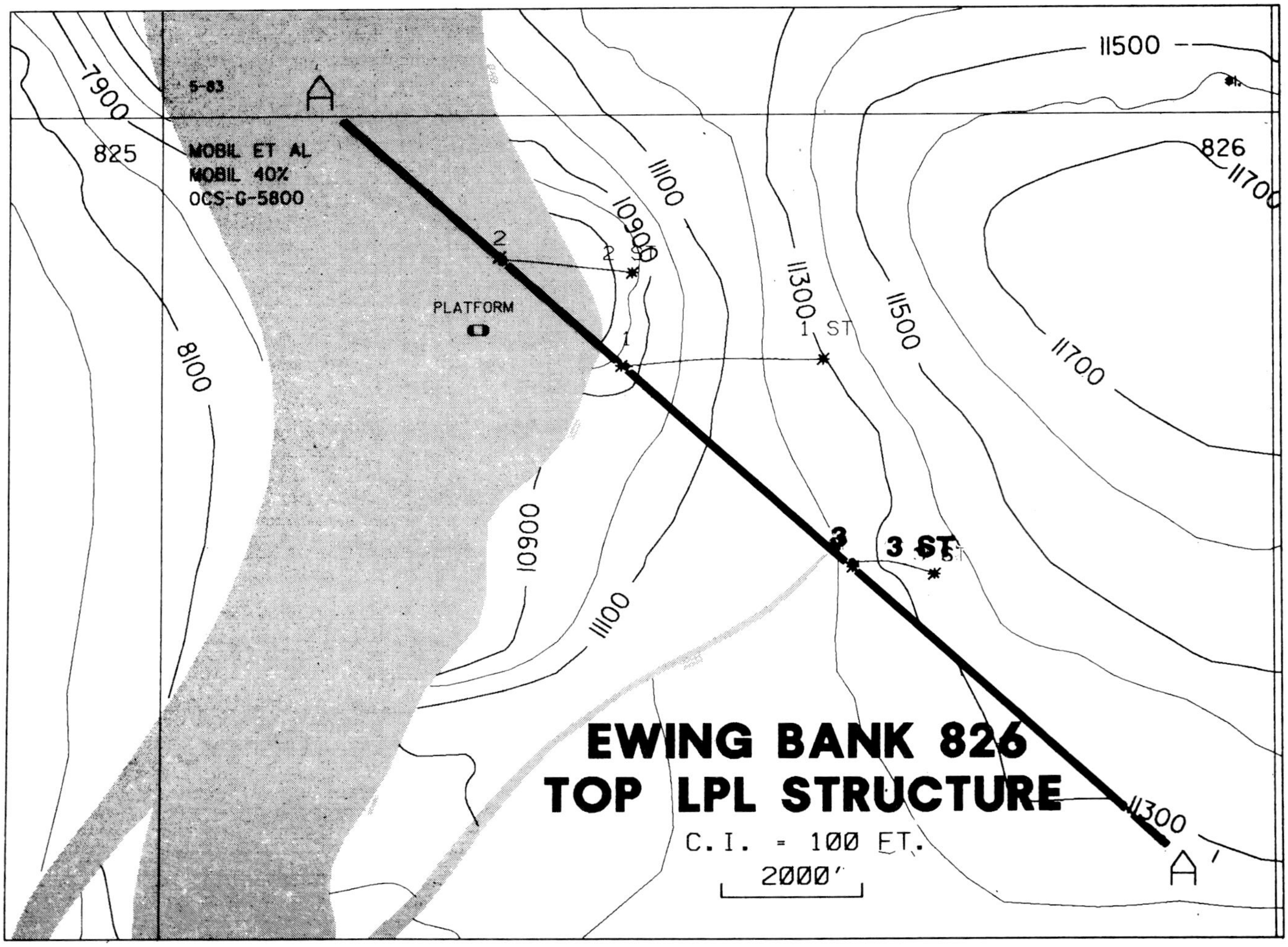

FIG. 15.—Map showing top LPL (Lower Pleistocene) structure. A cross section along A-A' is shown in Figure 16. Cores from wells 3 and 3ST are used in this study. Shaded zone represents area of growth fault.

August of 1989 because of high current velocities that reached nearly 3 knots (153 cm/sec) at 45 m depths, and 1 knot (51 cm/sec) at 250 m depths. These intense bottom currents affect the ability of a drilling rig to hold station over a wellhead (Koch et al., 1991).

Current measurements, bottom photographs of current ripples, high-resolution seismic records and GLORIA side-scan sonar records indicate that the Loop Current influences the sea floor at least periodically (Pequegnat, 1972; Bouma, 1972; Bouma and Hollister, 1973; Damuth et al., 1991). This is perhaps the most convincing evidence in demonstrating the existence of high energy of bottom currents in the Gulf of Mexico. Computed geostrophic flow velocities of the Loop Current vary from nearly 100 cm/sec at the sea surface to greater than 25 cm/sec at 500 m water depth (Nowlin and Hubert, 1972). Flow velocities measured using a current meter reach up to 19 cm/sec at a depth of 3286 m (Pequegnat, 1972). These currents are capable of reworking fine-grained sand. Pequegnat's (1972) bottom photographs of current ripples composed of sand at a depth of 3091 m (Alaminos Cruise 69-A-13, St.35) are an indication of deep bottom-current activity in the Gulf of Mexico today. These current ripples are draped by thin layers of mud. Such mud-draped ripples would produce mud-offshoots, as seen in the Ewing Bank cores (Fig. 10).

Mullins et al. (1987) suggested that the strong Loop Current/Gulf Stream circulation was created in the mid-Miocene due to closure of the Isthmus of Panama (Fig. 20). In addition, circulation of surface water masses may have been intensified during glacial periods of the Pliocene-

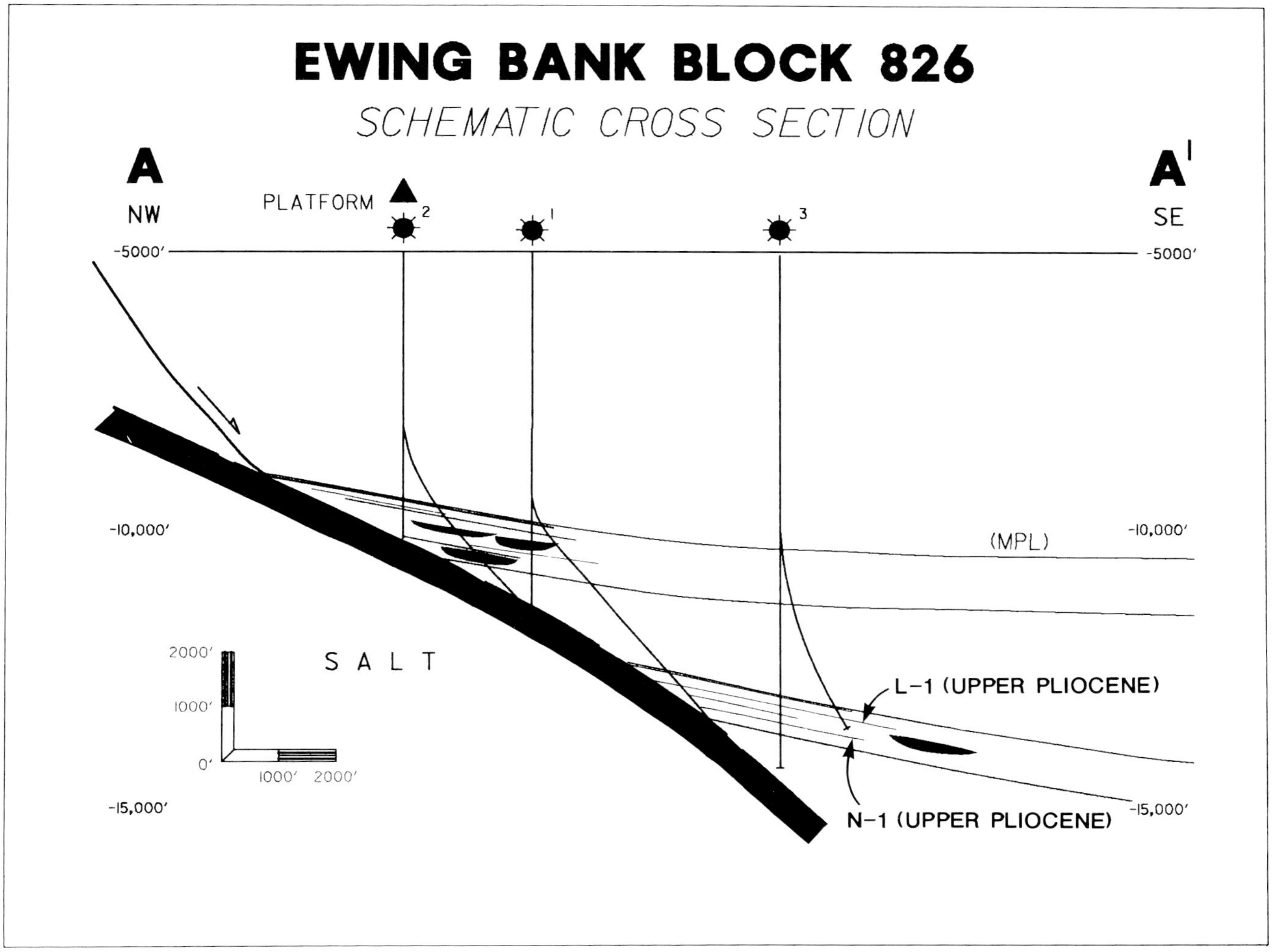

FIG. 16.—A schematic cross section across the Ewing Bank Block 826 Field showing stratigraphic position of Upper Pliocene L-1 and N-1 reservoir sands. The line of section- A-A'- is shown in Figure 15. MPL = Middle Pleistocene. Cores from both Upper Pliocene and Middle Pleistocene sequences were examined for this study.

Pleistocene (Arrhenius, 1952). Thus intensified Loop Current circulation may have led to sustained and stronger periods of current activity in the northern Gulf of Mexico. In support of this hypothesis, the Ewing Bank reservoirs tend to correlate with lowstands of sea level; similar to other reworked sands in the geologic record (Shanmugam and Moiola, 1982).

Depositional Model

We propose a depositional model for the Ewing Bank Block 826 area (Fig. 21) that involves the interaction of three types of currents, namely, channelized turbidity currents, overbank turbidity currents, and bottom currents. Facies encountered in the L-1 sand cored in the 3ST well consist of levee, turbidite channel, and bottom-current reworked sand intervals (Fig. 17), and is the basis for the lateral disposition in a depositional model (Fig. 21). Approximate position of the well 3 and well 3 sidetrack, their relative depths of penetration of individual reservoir sands L-1 and N-1, and cored intervals are illustrated in Figure 21. We suggest that the bottom-current reworked sands represent deposits of interchannel areas in a bathyal environment. Our model is not a conventional deep-sea fan developed by turbidity currents at the base-of-slope. Instead, the proposed model depicts a fine-grained facies

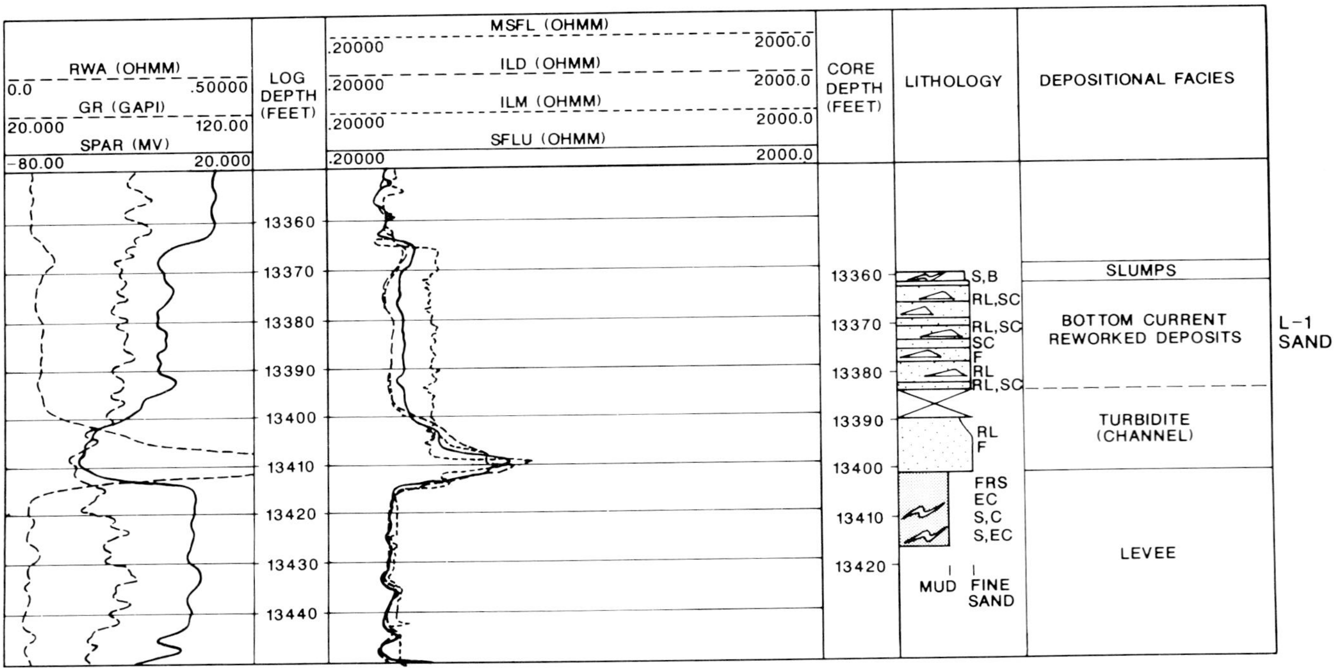

FIG. 17.—Vertical distribution of levee, turbidite channel, and bottom-current reworked facies and their log motifs in the L-1 sand. Well No. 3ST (Sidetrack), Ewing Bank Block 826, Gulf of Mexico. S = slump, B = breccia, RL = ripple cross-laminae, SC = sharp upper contact, F = fault, FRS = fracture-fill sand, EC = erosional contact, C = calcareous.

deposited from bottom currents as a distinctly different facies from channel-levee facies. Our bottom-current model has the potential for general application to other bathyal plays outside the study area with implications for reservoir distribution. For example, in deep-sea fans a levee facies commonly occurs in close association with channel facies. According to our model, in contrast, a seemingly similar-appearing fine-grained facies deposited by bottom currents may be present not in association with a channel facies. A major field of rippled sand is inferred to have developed on the western side of the channel-levee system. We suggest that an ancestral Loop Current was primarily responsible for the bottom-current reworking of these sands with perhaps minor influence from deep tidal currents.

Geometry of Sand Bodies

The lateral extent inferred for sand bodies in the the Ewing Bank 826 Field is based on both well and seismic data. In order to properly tie the Ewing Bank 826 No. 3 and No. 3ST wells to the 3-D seismic, two computer systems were used. A time-based cross-section along the deviated No. 3 well bore was created through MSAS, a VAX based Mobil software package on an Intergraph workstation. An arbitrary seismic line was also created at the same scale and orientation using a LANDMARK workstation. Figure 22 is an overlay of the time-based cross-section on the scaled arbitrary line from northwest to southeast showing amplitudes related to the L-1 and N-1 sands.

Although not penetrated in the sidetrack well, the N-1 sand is traceable in seismic sections beneath the sidetrack well (Fig. 22). The eastward transition of N-1 sand from bottom-current reworked facies into a channel/levee complex (see Fig. 21) is schematic. In general, the transition zone between bottom-current reworked sand and turbidite channel sand is composed of fine-grained levee facies. This facies association and related reservoir heterogeneity is critical in field development of bottom-current reworked sand reservoirs. There is no difference in seismic reflection patterns observed in the Ewing Bank 826 area between reworked sand and turbidite channel sand (see Figs. 21 and 22).

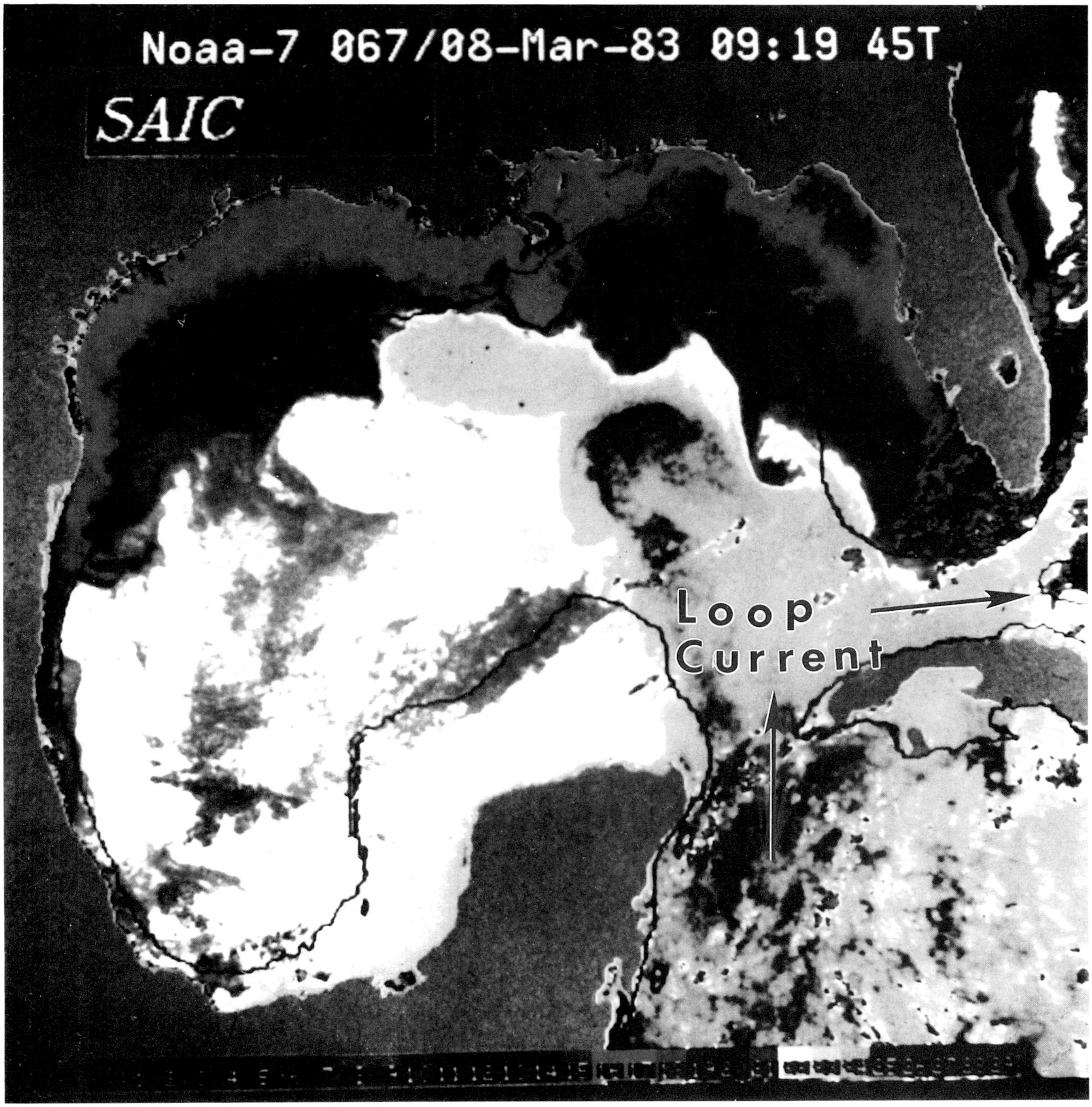

FIG. 18.—AVHRR (Advanced Very High Resolution Radiometry) satellite image of the Loop Current in the Gulf of Mexico. Courtesy of K. Luetkemeyer, University of Colorado at Boulder.

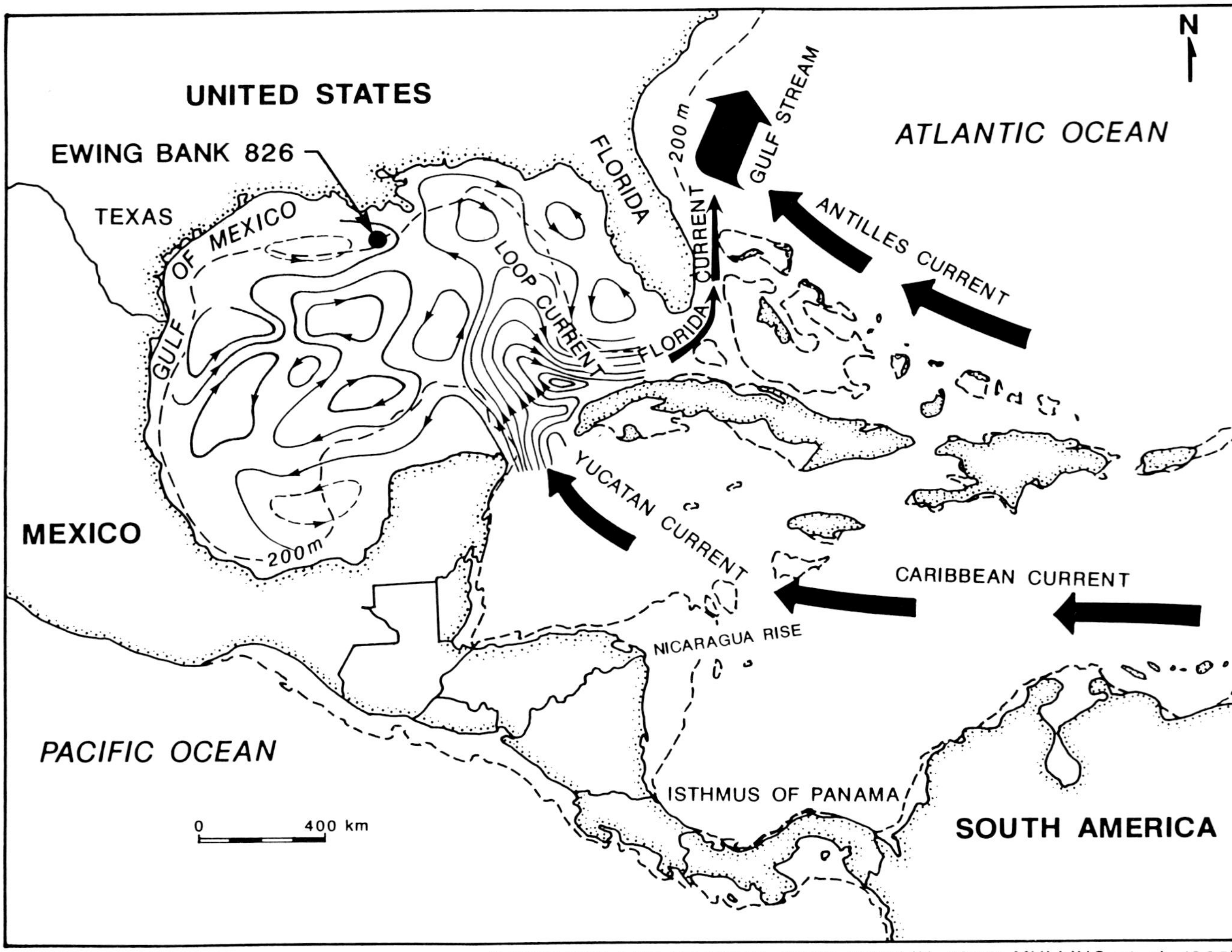

(NEUMANN AND PIERSON, 1966; NOWLIN, 1972; MULLINS et al, 1987)

FIG. 19.—Present circulation pattern of the Loop Current (after Neumann and Pierson, 1966; Nowlin, 1972; and Mullins et al, 1987). This wind-driven surface current is also considered to be affecting the sea bottom (Pequegnat, 1972). Note the eddies detached from the Loop Current in the Ewing Bank area.

In the absence of core, one could interpret the L-1 sand, showing continuous and parallel reflections in seismic profile (Fig. 22), as a laterally continuous sheet-like sand representing outer part of a submarine fan "lobe" and associated basin-plain facies (Shanmugam and Moiola, 1991). The core information, however, suggests that the L-1 sand is composed of a lower turbidite channel sand and an upper reworked sand (Fig. 21). This example underscores the need to integrate core information with seismic facies analysis in developing realistic depositional models. It is also critical to understand the origin of sands from core before correlating these "sands" over long distances using log motifs. This is because sands deposited by slumps, debris flows and bottom currents can generate similar log motifs, but the lateral continuity of sands can be quite variable.

On seismic profiles and well logs (Fig. 22) the L-1 sand appears to thicken toward the west suggesting a possible westerly source. Although not shown in the model (Fig. 21), a western channel-levee complex is inferred to have provided the sand for reworking by bottom currents. Subsequent salt movements in the area may have raised the L-1 sand causing an angular unconformity west of the No. 3 well due to erosion of the channel sands.

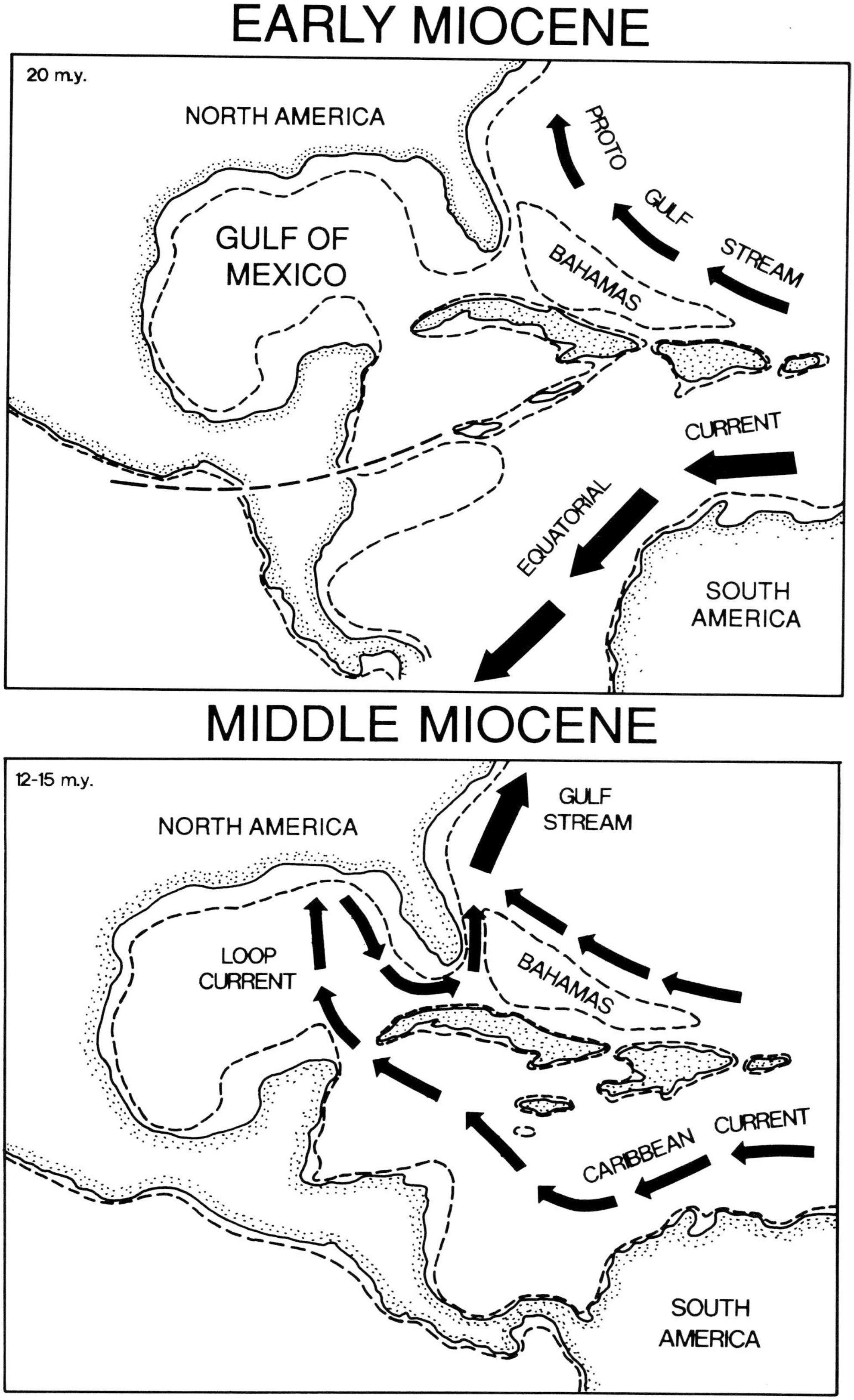

(MULLINS ET AL., 1987)

FIG. 20.—Changes in oceanic circulation patterns in the Gulf of Mexico since the Middle Miocene time due to the closure of Isthmus of Panama.

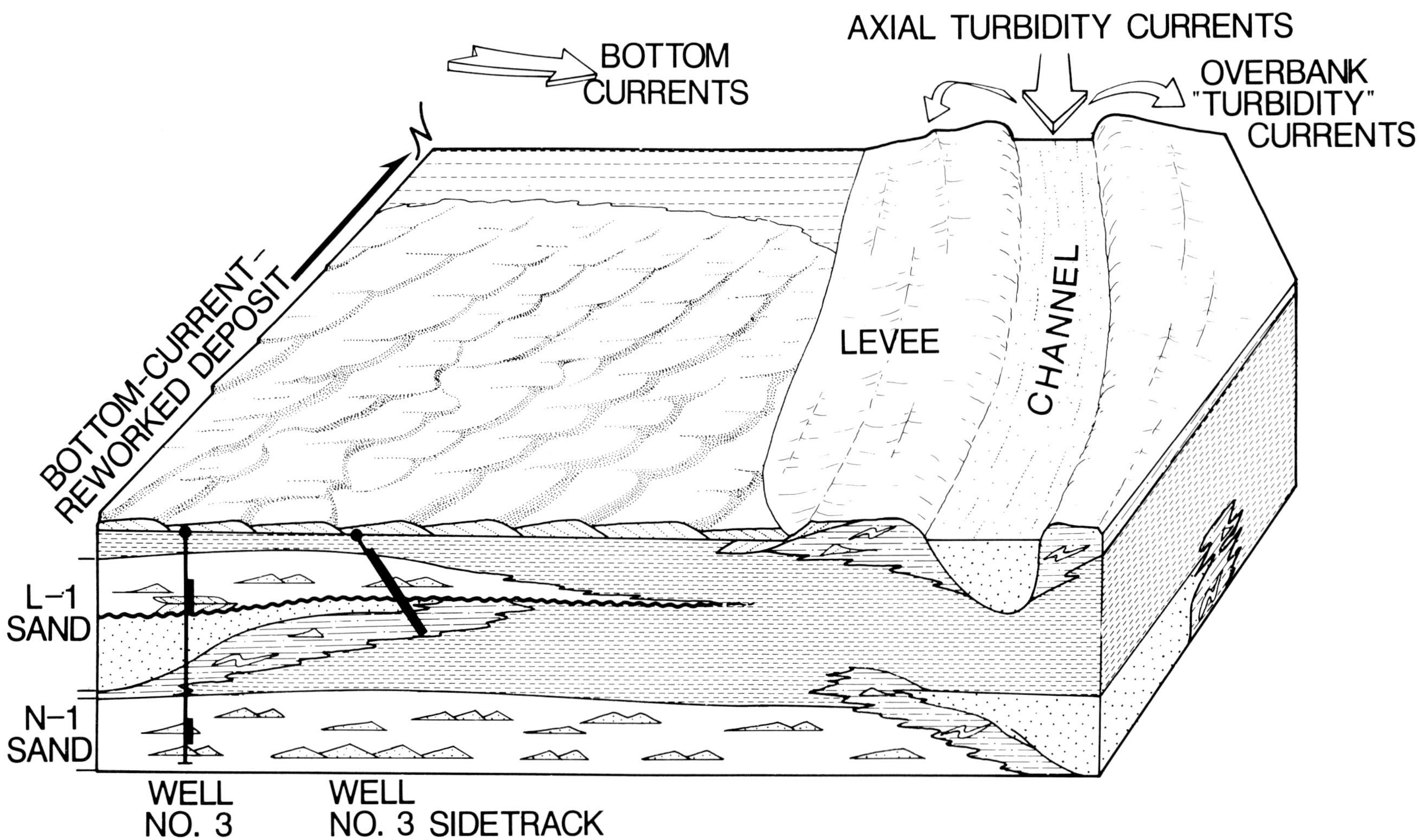

FIG. 21.—Schematic depositional model for the Pliocene and Pleistocene sand in the Ewing Bank Block 826 area. Note that bottom-current reworked sand in the interchannel area constitute a distinctly different facies from channel-levee facies. The model is based on vertical distribution of depositional facies in cored wells (see Fig. 17), log signatures, and seismic information (Fig. 22). Compare the position of No. 3, and No. 3ST wells in the model with their position in the seismic profile (Fig. 22). Cored intervals are shown by thick lines. Actual current directions of the Loop Current eddies are not shown for the study area.

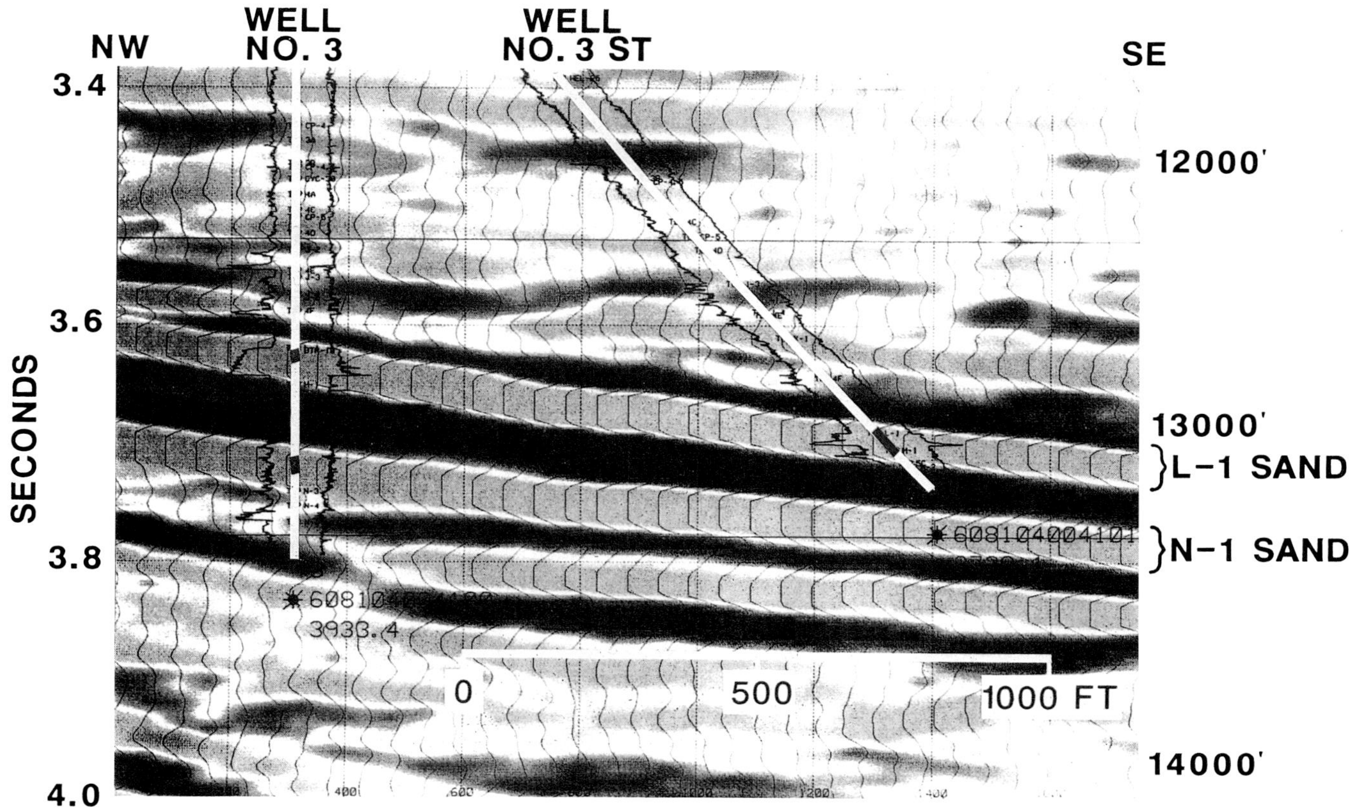

FIG. 22.—An arbitrary seismic profile showing position of cored wells through two sand units examined. Note continuous and parallel reflection pattern of the L-1 sand between the two wells, which may be interpreted to mean a tabular shape. However, core from the L-1 sand indicates the presence of a lower turbidite channel sand and an upper reworked sand (Fig. 21). Cored intervals are shown by black bars.

Stratigraphic dips of reworked sands (L-1) in Ewing Bank Block 826, processed from Stratigraphic High-resolution Dipmeter Tool (SHDT) data using the INCLINE program, exhibit consistent north-northwest stratigraphic dip. This paleocurrent direction is best explained by northward-flowing bottom currents caused by the detached eddies of the Loop Current (Fig. 19) rather than by turbidity currents that flow downslope in a southerly direction. Considering that the bulk of the cored interval of nearly 180 feet (55 m) in the Ewing Bank area represents primarily deposits of bottom currents, it is conceivable that this thick package may be laterally continuous in the study area (see geometry, Fig. 23), but individual sand layers within the package are thin and discontinuous (see core features, Fig. 23).

Reservoir Quality

Common ranges of porosity and permeability values, measured at ambient conditions, are given in Table 1. These values are expected to be lower at reservoir conditions. Both L-1 and N-1 sands in the No. 3 well show porosity values in the range of 25-40%. The horizontal air permeability, measured at 200 psi, commonly ranges from 100 to 1000 mD in the No. 3 well. Although the cored intervals of both L-1 and N-1 sands in the No. 3 well are products of bottom-current reworking, the L-1 reservoir with 80% sand exhibits higher permeability values (100-1800 mD) than the N-1 reservoir with 26% sand (50-800 mD). Microfractures, although seen in both L-1 and N-1 sands, enhance vertical permeability only in the L-1 sand

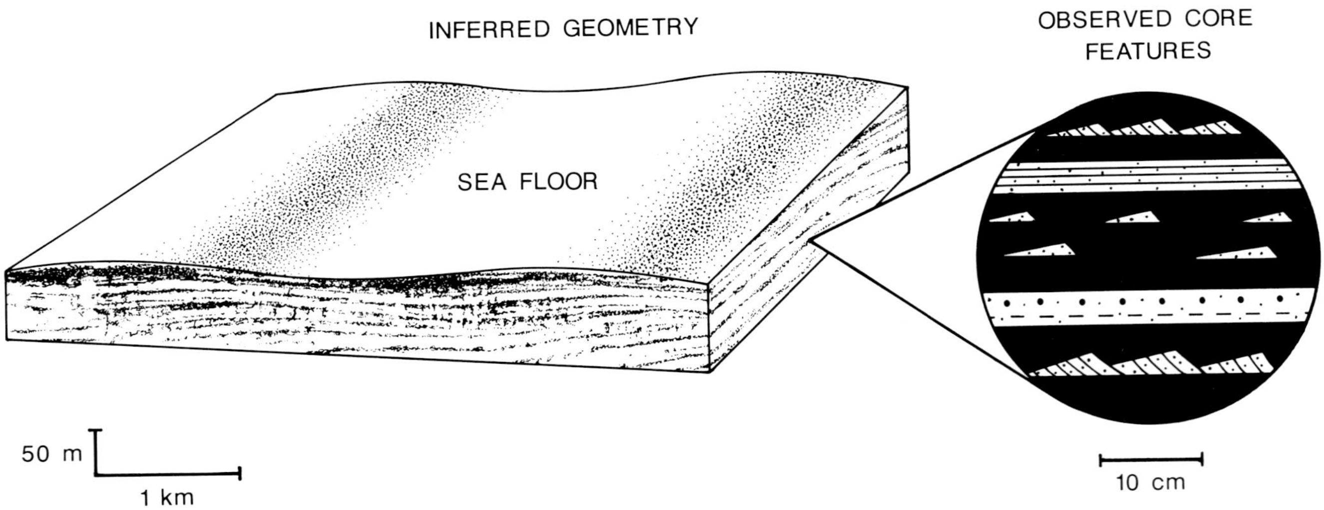

FIG. 23.—A conceptual model showing laterally continuous large-scale sediment body composed of small-scale discontinuous sand layers.

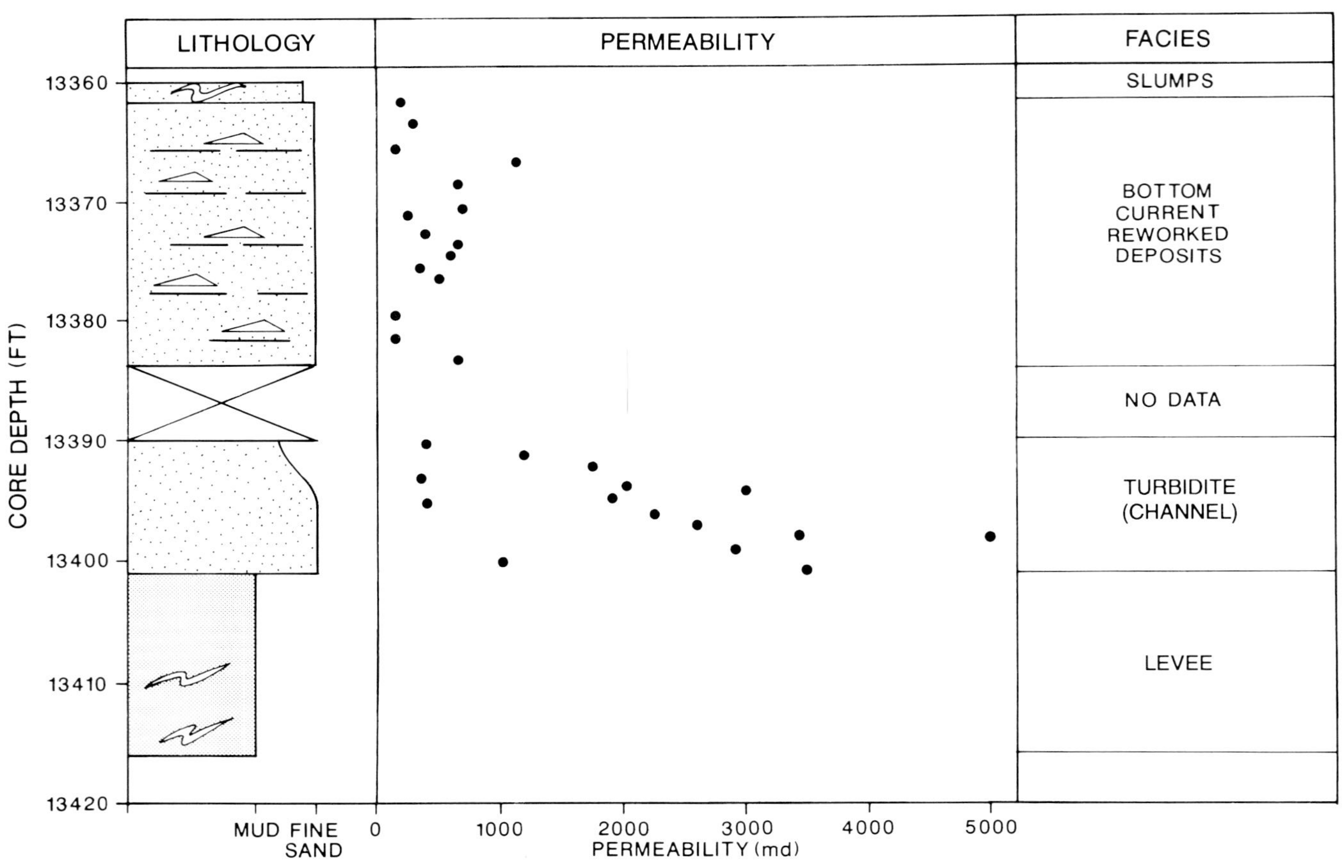

FIG. 24.—Permeability distribution in the L-1 Sand. Note upward decrease in permeability within the turbidite channel unit reflecting upward decrease in grain size. In comparison, permeability within bottom-current reworked facies is much less variable because of uniform grain size. Lower permeability values of bottom-current reworked facies may be attributed to interbedded mud. Well No. 3ST, Ewing Bank Block 826, Gulf of Mexico.

TABLE 1.— COMMON RANGE OF POROSITY AND PERMEABILITY VALUES (MEASURED FROM CORE PLUGS AT 200 PSI).

because of higher sand content (80%), which allows a larger number of sand on sand contacts. The vertical permeability values of N-1 sand are expected to be poor because of the ubiquitous presence of interbedded mud throughout the sequence. As a result, sand layers commonly offset against mud layers across fractures in the N-1 sand. We also speculate that the transitional, fine-grained, levee facies that occurs between channel sand and reworked sand (Fig. 21) may exhibit low permeability values within the N-1 reservoir.

Although both turbidite and reworked sands show porosity values in the range of 35-42%, their permeability values are strikingly different. In the No. 3ST well, for example, the basal turbidite channel sand unit shows a distinct upward-decrease in permeability with maximum values (5000 mD) at its base (Fig. 24). The permeability trend perhaps mimics the subtle upward-decrease in grain size within the channel package. The upper unit with bottom-current-reworked deposits exhibits a near uniform vertical distribution of permeability values of approximately 500 mD (Fig. 24), reflecting uniform vertical grain size distribution.

On the basis of reservoir quality, sand in the Ewing Bank area can be classified into two types (Fig. 25). The Type 1 exhibits a higher number of sand layers per unit area than the Type 2. The permeability of Type 1 (L-1) is much higher (up to 1800 mD) than in Type 2 (N-1) (up to 800 mD). The porosity values, however, remain the same for both (Fig. 25). We attribute the higher permeability in the Type 1 to: (1) increased overall sand content, (2) vigorous bottom-current reworking, and (3) microfractures.

The Type 1 sands are clean, porous, and well sorted. Close vertical stacking of sand layers and a lack of interbedded mud layers facilitate good communication between sand. The communication between sand layers is further enhanced by fractures. For these reasons, the Type 1 sands should produce at good levels. In contrast, the Type 2 sands contain numerous interbedded mud layers, which greatly reduce communication between sand layers. Therefore, their producibility is expected to be poor. In fact, the Type 1 sands with good communication between sand layers have produced at higher rates and recovery efficiencies than the Type 2 sands.

SUMMARY

Sedimentary structures of Pliocene-Pleistocene cores from the Ewing Bank Block 826 area in the Gulf of Mexico indicate reworking by bottom-currents at bathyal water depths. Bottom-current reworked sands (sandy contourites) are characterized by traction bedforms, such as cross-bedding, horizontal lamination, starved and isolated current ripples, sharp upper contacts, and inverse grading. An important attribute of reworked sands is that the traction bedforms occur in discrete beds, and are not part of a regular vertical sequence, such as the Bouma sequence that is formed by waning turbidity flows. The reworked sands also exhibit internal erosional surfaces indicating pulses of increased current energy, and mud-offshoots suggesting oscillating energy conditions.

The Loop current, a strong wind-driven surface current in the Gulf of Mexico, impinged on the sea bottom, and reworked sand. A depositional model was developed for the reworked sands by integrating well (core and log) and 3-D seismic data. In this model, bottom-current reworked sands tend to occur in an interchannel area. The reworked sediment package may be thick and continuous, but individual sand layers within the package may be thin and discontinuous.

Although the cored intervals of both L-1 (Type 1) and N-1 (Type 2) reservoir sand in the No. 3 well in Ewing Bank show similar porosity values (25-40%), the L-1 reservoir with 80% sand exhibits higher permeability values (100-1800 mD) than the N-1 reservoir with 26% sand (50-800 mD). The lower permeability values of the N-1 sand can be explained by the presence of numerous

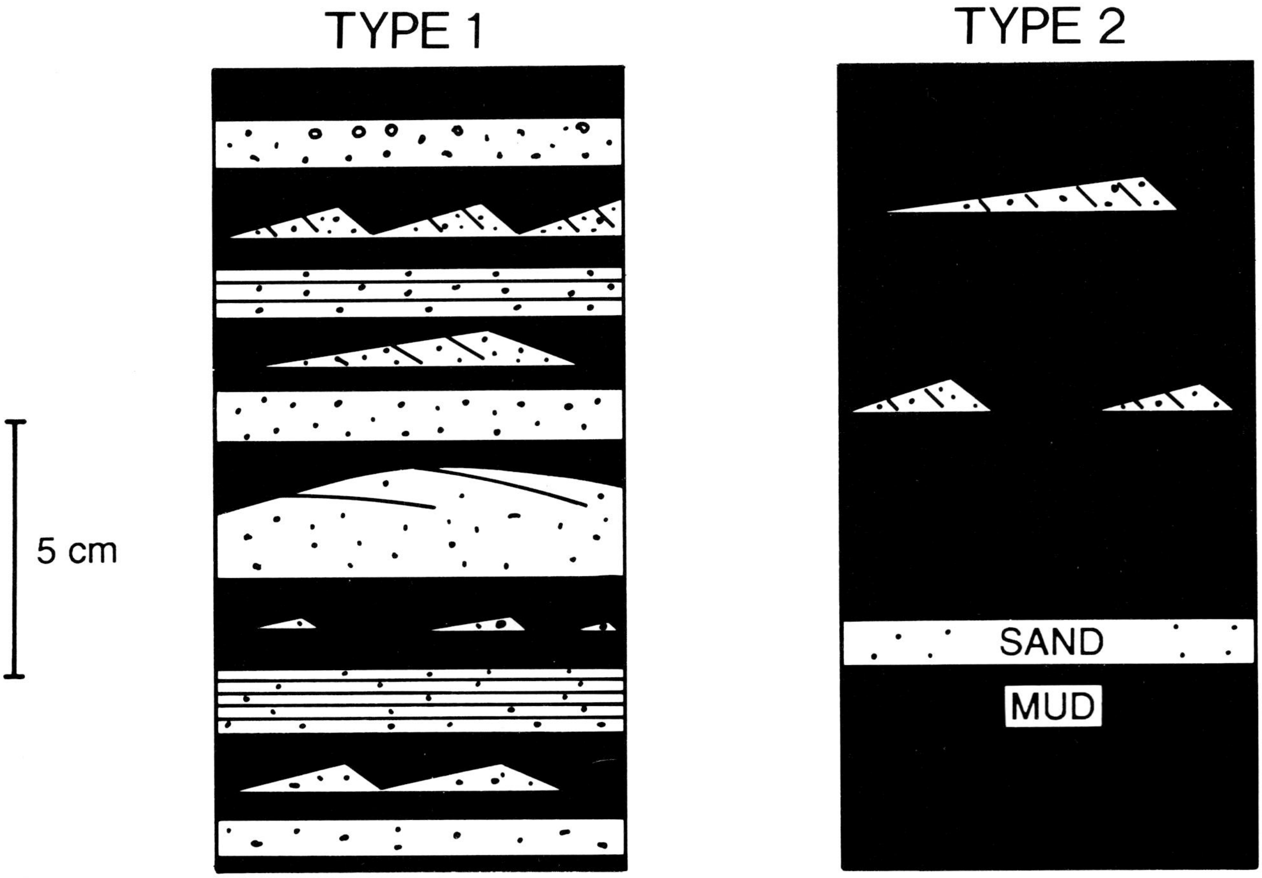

	TYPE 1	TYPE 2
SAND	80%	25%
POROSITY	25-40%	25-40%
PERMEABILITY	100-1800 MD	50-800 MD
UNIT	L1	N1

FIG. 25.—Two types of reservoir facies showing differences in sand content and permeability. Type 1 with higher sand content exhibits higher permeability than sand-poor Type 2. Note both types show similar values of porosity from core plugs taken from the sand layers.

interbedded mud layers throughout the sequence. The producibility of L-1 (Type 1) sand is better than the N-1 (Type 2) sand.

ACKNOWLEDGEMENTS

We thank R. D. Winn for his critical and thorough review of the manuscript; J. A. Armentrout for editorial help; K. Luetkemeyer (University of Colorado at Boulder) for providing an AVHRR satellite image of the Loop Current; G.K. Baker for managerial support; M. K. Lindsey and A. Long for drafting; and S. Kizer for reproducing illustrations. We are thankful to Mobil and its partners BP Exploration and Kerr-McGee for granting permission to publish this paper.

REFERENCES

ALLEN, J. R. L., 1982, Sedimentary Structures, Their Character and Physical Basis, v. 1: Amsterdam, Elsevier, 593 p.

ANSPACH, D. H., TRIPP, S. E., BERLITZ, R. E., AND GILREATH, J. A., 1989, Study of shelf-slope environments of deposition, High Island A-474/A-499 Area (abs.): American Association of Petroleum Geologists Bulletin, v. 73, p. 328-329.

ARRHENIUS, G., 1952, Sediment cores from the east Pacific: Reports of the Swedish Deep-Sea Expedition 1947-1948, (1), 89 p.

BERGGREN, W.A., AND HOLLISTER, C. D., 1977, Plate tectonics and paleocirculation - Commotion in the ocean: Tectonophysics, v. 38, p. 11-48.

BOUMA, A. H., 1962, Sedimentology of some flysch deposits: A graphic approach to facies interpretation: Amsterdam, Elsevier, 168 p.

BOUMA,, A. H., 1972, Recent and ancient turbidites and contourites: Transactions - Gulf Coast Association of Geological Societies, v. 22, p. 205-221.

BOUMA, A. H., AND HOLLISTER, C. D., 1973, Deep ocean basin sedimentation, *in* Middleton, G. V., and Bouma, A. H., eds., Turbidites and deep-water sedimentation: Los Angeles, California, Society of Economic Paleontologists and Mineralogists Pacific Section Short Course: p. 79-118.

CHOUGH, S. K., AND HESSE, R., 1985, Contourites from the Eirik Ridge, South of Greenland: Sedimentary Geology, v. 41, p. 185-199.

COOPER, C., FORRISTALL, G.Z., AND JOYCE,T.M., 1990, Velocity and hydrographic structure of two Gulf of Mexico warm-core rings: Journal of Geophysical Research, v. 95, No. C2, p. 1663-1679.

DAMUTH, J. E. ET AL., 1991, Neogene depositional processes on the Texas-Louisiana continental slope revealed by high-resolution seismic studies: American Association of Petroleum Geologists Annual Convention Official Program, abs. p. 94-95.

ENJOLRAS, J. M., GOUADAIN, J., MUTTI, E., ANDPIZON, J., 1986, New turbiditic model for the Lower Tertiary sands in the South Viking Graben, *in* Spencer, A. M., et al., eds., Habitat of Hydrocarbons on the Norwegian Continental Shelf: London, Graham and Trotman, p. 171-178.

EWING, M., EITTREIM, S., EWING, J., AND LE PICHON, X., 1971, Sediment transport and distribution in the Argentine Basin, 3. Nepheloid layer and processes of sedimentation: Physics and Chemistry of the Earth, v. 8, p. 49-78.

FLOOD, R. D., AND HOLLISTER, C. D., 1974, Current-controlled topography on the continental margin off the eastern United States, *in* Burke, C. A., and Drake, C. L., eds., The Geology of Continental Margins, New York, Springer-Verlag, p. 197-205.

FLOOD, R. D., AND SHOR, A. N., 1988, Mud waves in the Argentine Basin and their relationship to regional bottom circulation patterns: Deep-Sea Research, v. 35, p. 943-971.

GONTHIER, E. G., FAUGERES, J.-C., AND STOW, D. A. V., 1984, Contourite facies of the Faro Drift, Gulf of Cadiz, *in* Stow, D. A. V. , and Piper, D. J. W., eds., Fine-Grained Sediments: Deep-Water Processes and Facies: Geological Society of London Special Publication 15, p. 275-292.

HEEZEN, B. C., HOLLISTER, C. D., AND RUDDIMAN, W. F., 1966, Shaping of the continental rise by deep geostrophic contour currents: Science, v. 152, p. 502-508.

HOLLISTER, C. D., 1967, Sediment distribution and deep circulation in the western North Atlantic: Unpublished Ph.D. dissertation, Columbia University, New York, 467 p.

HOLLISTER, C. D., AND HEEZEN, B. C., 1972, Geologic effects of ocean bottom currents: western north Atlantic, *in* Gordon, A. L., ed., Studies in physical oceanography, volume 2: New York, Gordon and Breach Science Publishers, p. 37-66.

HOLLISTER, C. D., AND MCCAVE, I. N., 1984, Sedimentation under deep-sea storms: Nature, V. 309, p. 220-225.

HUBERT, J. F., 1964, Textural evidence for deposition of many western North Atlantic deep-sea sands by ocean-bottom currents rather than turbidity currents: Journal of Geology, v. 72, p. 757-785.

KARL, H. A., CACCHIONE, D. A., AND CARLSON, P. R., 1986, Internal-wave currents as a mechanism to account for large sand waves in Navarinsky Canyon Head, Bering Sea: Journal of Sedimentary Petrology, v. 56, p. 706-714.

KOCH, S. P., BARKER, J. W., AND VERMERSCH, J. A., 1991, The Gulf of Mexico Loop Current and deepwater drilling: Journal of Petroleum Technology, v. 43, p. 1046-1050 and 1118-1119.

LOVELL, J. P. B., ANDSTOW, D. A. V., 1981, Identification of ancient sandy contourites: Geology, v. 9, p. 347-349.

LOWE, D. R., 1979, Sediment gravity flows: Their classification, and some problems of application to natural flows and deposits: SEPM Special Publication 27, p. 75-82.

MITCHUM, JR., R.M., ET AL., 1990, Sequence stratigraphy in Late Cenozoic expanded sections, Gulf of Mexico: 11th Annual Research Conference, Gulf Coast SEPM Foundation, p. 237-256.

MULLINS, H. T., ET AL., 1987, Middle Miocene oceanographic event in the eastern Gulf of Mexico: Implications for seismic stratigraphic succession and Loop Current/Gulf Stream circulation: Geological Society of America Bulletin, v. 98, p. 702-713.

MUTTI, E., 1977, Distinctive thin-bedded turbidite facies and related depositional environments in the Eocene Hecho Group (southcentral Pyrenees, Spain): Sedimentology, v. 24, p. 107-131.

MUTTI, E., 1992, Turbidite sandstones: Milan, Italy, Agip, 275 p.

MUTTI, E., PAROOS, N., POFADO, S., AND RUMENOS, L., 1980, Deep sea fan turbidite sediments winnowed by bottom currents in the Eocene of the Campos Basin- Brazilian offshore: Ist European Meeting International Association of Sedimentologists, Abstracts, p. 114.

NEUMANN, G., AND PIERSON, W. J., JR., 1966, Principles of Physical Oceanography: Englewood Cliffs, New Jersey, Prentice-Hall, 545 p.

NOWELL, A. R. M., MCCAVE, I. N., AND HOLLISTER, C. D., 1985, Contributions of HEBBLE to understanding of marine sedimentation: Marine Geology, v. 66, p. 397-409.

NOWLIN, W. D., JR., 1972, Winter circulation patterns and property distributions: *in* Capurro, L. R. A., and Reid, J. L., eds., Contributions on the Physical Oceanography of the Gulf of Mexico: Texas A&M University Oceanographic Studies, Volume 2: Houston, Gulf Publishing Co., p. 3-51.

NOWLIN, W. D., JR., AND J. M. HUBERT, 1972, Contrasting summer circulation patterns for the eastern Gulf, *in* Capurro, L. R. A., and Reid, J. L., eds., Contributions on the Physical Oceanography of the Gulf of Mexico: Texas A&M University Oceanographic Studies, Volume 2: Houston, Gulf Publishing Co., p. 119-137.

PEQUEGNAT, W. E., 1972, A deep bottom current on the Mississippi Cone: *in* Capurro, L. R. A., and Reid, J. L., eds., Contributions on the Physical Oceanography of the Gulf of Mexico: Texas A&M University Oceanographic Studies, Volume 2: Houston, Gulf Publishing Co., p. 65-87.

PIPER, D. J. W., AND BRISCO, C. D., 1975, Deep-water continental-margin sedimentation, DSDP Leg 28, Antarctica, *in* Hayes, D. E., et al., Initial Reports of the Deep Sea Drilling Project: Washington, D.C., U. S. Govt. Printing Office, p. 727-755.

REINECK, H. E., AND SINGH, I. B., 1980, Depositional Sedimentary Environments, Second Edition: New York, Springer-Verlag, 549 p.

RICHARDSON, M. WIMBUSH, J., M., AND MAYER, L., 1981, Exceptionally strong near-bottom flows on the continental rise of Nova Scotia: Science, v. 213, p. 887-888.

SHANMUGAM, G., AND MOIOLA, R. J. , 1982, Eustatic control of turbidites and winnowed turbidites: Geology, v. 10, p. 231-235.

SHANMUGAM, G. ET AL., 1988, Comparison of turbidite facies associations in modern passive-margin Mississippi fan with ancient active-margin fans: Sedimentary Geology, v. 58, p. 63-77.

SHANMUGAM, G., 1990, Deep-marine facies models and the interrelationship of depositional components in time and space, *in* Brown, G.C., Gorsline, D.S., and Schweller, W.J., eds., Deep-Marine Sedimentation: Depositional models and case histories in hydrocarbon exploration and development: San Francisco, Society of Economic Paleontologists and Mineralogists Pacific Section, p. 199-246.

SHANMUGAM, G., SPALDING, T.D., AND ROFHEART, D.H., 1993a, Process sedimentology and reservoir quality of deep-marine bottom-current reworked sands (sandy contourites): An example from the Gulf of Mexico: American Association of Petroleum Geologists Bulletin, v. 77, 1241-1259.

SHANMUGAM, G., SPALDING, T.D., AND ROFHEART, D.H., 1993b, Traction structures in deep-marine bottom-current reworked sands in the Pliocene and Pleistocene, Gulf of Mexico: Geology, v. 21, p. 929-932.

SHANMUGAM, G., AND MOIOLA, R. J., 1991, Types of submarine fan lobes: models and implications: American Association of Petroleum Geologists Bulletin, v. 75, p. 156-179.

SHEPARD, F. P., ET AL., 1979, Currents in submarine canyons and other sea valleys: AAPG Studies in Geology 8, 173 p.

STOW, D. A. V., AND J. BOWEN, A., 1978, Origin of lamination in deep sea, fine-grained sediments: Nature, v. 274, p. 324-328.

STOW, D. A. V., AND LOVELL, J. P. B., 1979, Contourites: Their recognition in modern and ancient sediments: Earth-Science Reviews, v. 14, p. 251-291.

STOW, D. A. V., AND HOLBROOK, J. A., 1984, North Atlantic contourites: an overview, *in* Stow, D. A. V., and Piper, D. J. W., eds, Fine-Grained Sediments: Deep-Water Processes and Facies: Geological Society Special Publication No. 15, p. 245-256.

STOW, D. A. V., AND SHANMUGAM, G., 1980, Sequence of structures in fine-grained turbidites: Comparison of recent deep-sea and ancient flysch sediments: Sedimentary Geology, v. 25, pp. 23-42.

STOW, D.A.V., AND FAUGERES, J.C., eds., 1993, Contourites and bottom currents: Sedimentary Geology, v. 82, special volume, 310 p.

TUCHOLKE, B. E., AND EMBLEY, R. W., 1984, Cenozoic regional erosion of the abyssal sea floor off South Africa, *in* Schlee, J. S., ed., Interregional Unconformities and Hydrocarbon Accumulation: American Association of Petroleum Geologists Memoir 36, p. 145-164.

TUCHOLKE, B. E., HOLLISTER, C. D., BISCAYE, P. E., AND GARDNER, W. D., 1985, Abyssal current character determined from sediment bedforms on the Nova Scotian continental rise: Marine Geology, v. 66, p. 43-57.

UNRUG, R., 1977, Ancient deep-sea traction currents deposits in the Lgota Beds (Albian) of the Carpathian flysch: Rocznik Polskiego Towarzystwa Geologicznego Annales de la Societe Geologique de Pologne, v. 47, p. 355-370.

VAIL, P. R., AND WORNARDT, W. W., 1990, Well log-seismic sequence stratigraphy: An integrated tool for the 90's, *in* Sequence Stratigraphy as an Exploration Tool, Concepts and Practices in the Gulf Coast: 11th Annual Research Conference, Gulf Coast SEPM Foundation, p. 379-388.

ZHENZHONG, G., AND ERIKSSON, K. A., 1991, Internal-tide deposits in an Ordovician submarine channel: Previously unrecognized facies?: Geology, v. 19, p. 734-737.

GREEN CANYON 184 UPPER SLOPE TURBIDITES

WILLIAM J. SCHNEIDER
H. EDWARD CLIFTON
Conoco, Inc., Lafayette, LA 70508
Conoco, Inc. Houston, TX 77252

ABSTRACT

Jolliet Field reservoirs, discovered 1981, are middle to upper Pleistocene intraslope turbidites complicated by rapid deposition, sea level fluctuations, salt movement, and structural complexities. Seven exploratory wells integrated with 3-d seismic coverage allowed for geological, geophysical, and petrophysical description of the reservoirs. Subsequent drilling of 20 development wells provided additional data and prompted the recognition of a significant thrust fault.

Production began from the field in November, 1989. High initial rates with rapid declines were characteristic of the reservoirs, especially in the subthrust and faulted east flank. This behavior is attributed to reservoir compartmentalization and lack of aquifer support. Completions in the less faulted western flank have been much better performers. Produced oils have ranged 20 to 60 degrees in API gravity, and some oils are paraffinic.

Notable technical successes in the field include the predrilling of 20 development wells under time and under budget, the installation of the first tension leg well platform in the Gulf of Mexico in what was at the time a water depth record for a production platform (1760 feet), gravel packs of fine grained zones, and the use of selective completions. Significant concerns include the poor performance of some reservoirs and the production of paraffin.

Core from two wells in the field provide insight into the depositional processes responsible for the accumulation of two of the reservoir sands, the HJ sand and the KE sand. The HJ sand was cored in Green Canyon well 184 #7. The unit shows a succession of lithofacies that are consistent with development from either a small, migrating, channel-levee-overbank system or a prograding channel/lobe system. A succession of similar scale in a Paleocene submarine canyon fill exposed at Point Lobos, California, provides a possible analogue. Abundant graded beds and Bouma sequences indicate that the uppermost and lowermost parts of the HJ sand were deposited predominantly from turbidity currents, and an intensely ripple-laminated middle interval represents reworking either at a hydraulic jump or by contour currents. In contrast, the KE sand seems more to have resulted from debris flows, at least in its occurrence in the Green Canyon well 184 A-1. Mud-clast breccia or conglomerate is common throughout the core, and the sand beds are more massive and lack Bouma sequences. In the Green Canyon 184 #7 sidetrack well, the KE sands are better stratified and may have been more influenced by deposition from turbidity currents.

INTRODUCTION

Green Canyon 184 was acquired in 1980 for $7,800,000 by a consortium of Conoco, Cities Service, and Getty which later became Conoco, Oxy, and Texaco. The block is located 125 miles offshore Louisiana (Figure 1) in water depths of 900 to over 2000 feet. Drilled in 1981, it was the first Gulf of Mexico discovery in greater than 1000 feet of water. Subsequent drilling proved up reserves in the upper and middle Pleistocene slope turbidites on this complexly faulted anticline. A template was set in 1760 feet of water, 20 production wells were predrilled, the world's first

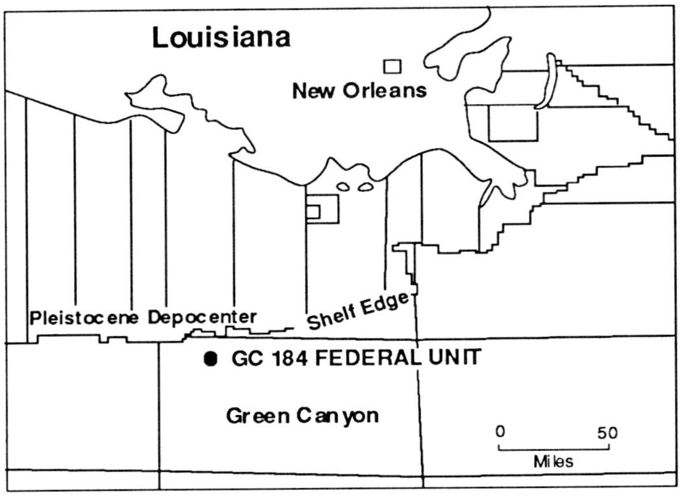

FIG. 1.—Map of offshore Louisiana showing the location of Green Canyon Block 184.

tension-leg well platform (TLWP) was installed, and the wells then completed. First oil was produced in November, 1989, from what is now called Jolliet Field.

STRUCTURAL SETTING

Green Canyon 184 lies within a present day upper slope structurally dominated by salt diapirs and layers as shown in Figure 2. The Jolliet field structural trap is formed by an anticline complexly faulted by both normal and reverse faults located on the western edge of a major salt withdrawal basin. Mapping with 3-D seismic data combined with the drilling of the development wells documented the existence of a major thrust fault (Cook and D'Onfro, 1991). The thrust fault is shown on the east edge of the structure in Figure 3 and in cross-section in Figure 4. The thrust fault appears related to movement along a decollement formed by a salt layer with the majority of movement towards the south-southeast. Subsequent to and concurrent with the thrust fault the field has been cut by a series of roughly north-south down to the east normal faults (Figures 3 and 4). In addition to the mapped faults there appear to be more faults that are below seismic resolution, especially on the east flank and subthrust areas where production histories show limited drainage areas.

STRATIGRAPHIC SETTING

Reservoirs in Jolliet field are middle to upper Pleistocene in age. Figure 5 shows the relationship of the principle pay zones to key paleontologic markers as well as to fluctuations in sea level. Principle pays are located between *Trimosina A* and *Angulogerina B*. In general individual beds recognized in core or on high resolution dipmeters are less than six inches thick with some in channel settings with an aggregate thickness of more than ten feet. Sand depositional units average 100 feet thick with a 50 to 60 percent net to gross ratio. The sandstones classify as sub-litharenites and contain 60 to 70 percent quartz, 15 to 20 percent feldspar, and 10 to 20 percent mostly calcitic rock fragments. Clays are mixed layer illite/smectite with minor kaolinite and chlorite. High feldspar content and thin bedding with shales makes log interpretation a challenge.

The sediments are interpreted to be upper slope turbidites with some reworking due to slumping and bottom currents. The elements of the depositional environment are a shelf edge delta as evidenced by paleontologic control and a depositional maximum located 20 to 30 miles to the north in Eugene Island South Addition, from which sands fed into the upper slope and accumulated in the narrow salt-controlled basins. During times of major canyon development, sediment bypassed the area either laterally or down dip. Figure 6 is a cross-section which shows some of the amalgamated channel-levee sands of the I and H series sands. While the package as a whole is easily traceable over the field, individual sands are often traceable only 500 to 1500 feet laterally. Some water contacts high on structure may be due to water trapped stratigraphically, and current production data indicates that while some zones are locally well connected others are not.

DEVELOPMENT

Seven exploratory wells were drilled between 1981 and 1985, culminating in the Green Canyon 184 #7 which found 606 feet net effective pay spread over 22 sands. Development was approved by the partners in 1986. The development involved predrilling 20 wells through a 24 slot template using a semi-submersible drill rig, setting a tension leg well platform (TLWP) and completing the wells with a small rig on the TLWP. Gas and oil-water emulsion are produced through separate pipelines to a central processing platform 12 miles to the north in Green Canyon 52. This facility also serves Marquette field located in Green Canyon 52.

Predrilling of the production wells was accomplished under the allotted time and under budget. These wells

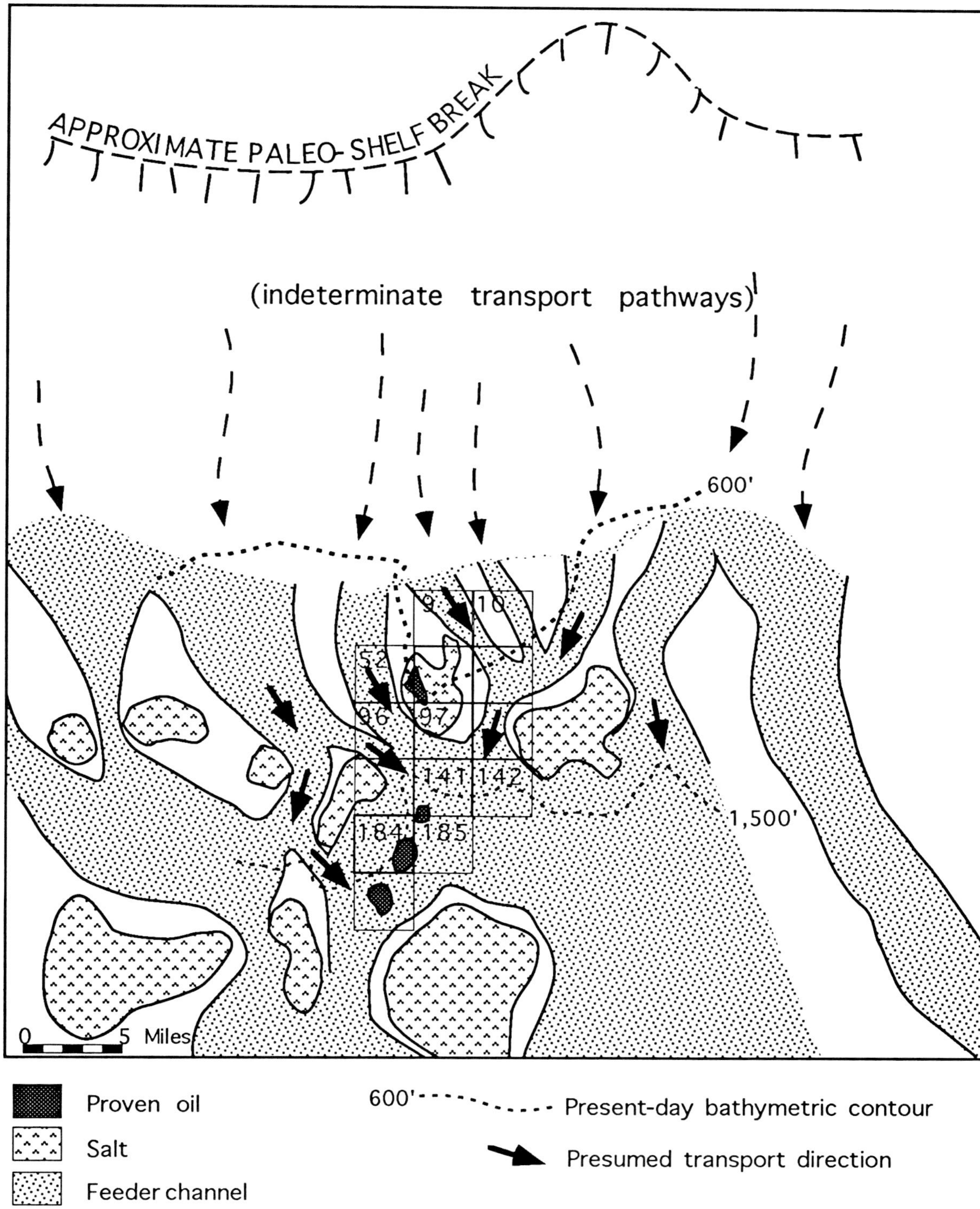

Fig. 2.—Map of the Green Canyon Federal Unit, showing salt diapirs, feeder channels, proven oil locations, and approximate position of the shelf break.

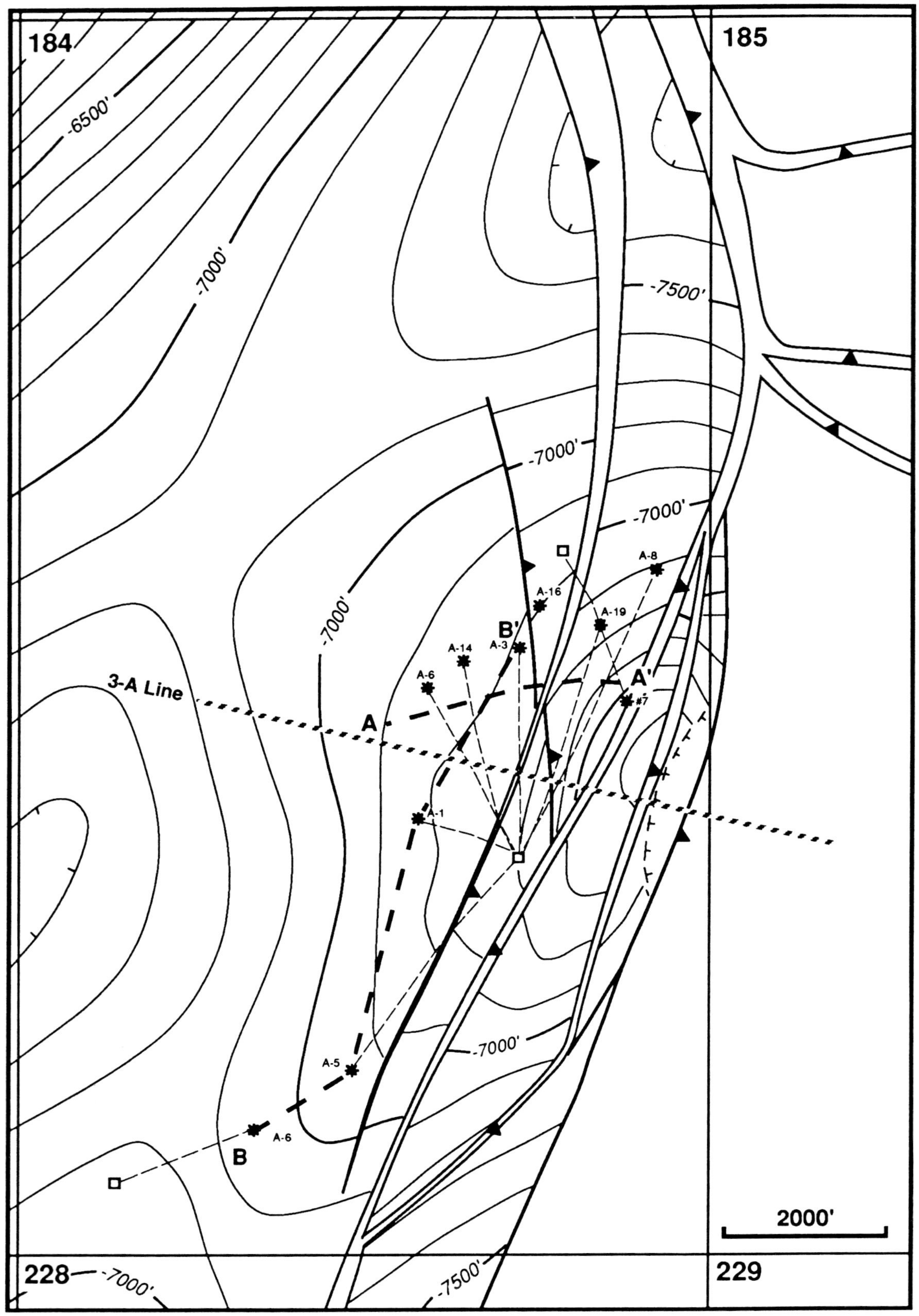

Fig. 3.—Top of the "HO" sand in Green Canyon Block 184 , based on 3-D seismic data. From Doug Cook, 1989, internal Conoco report.

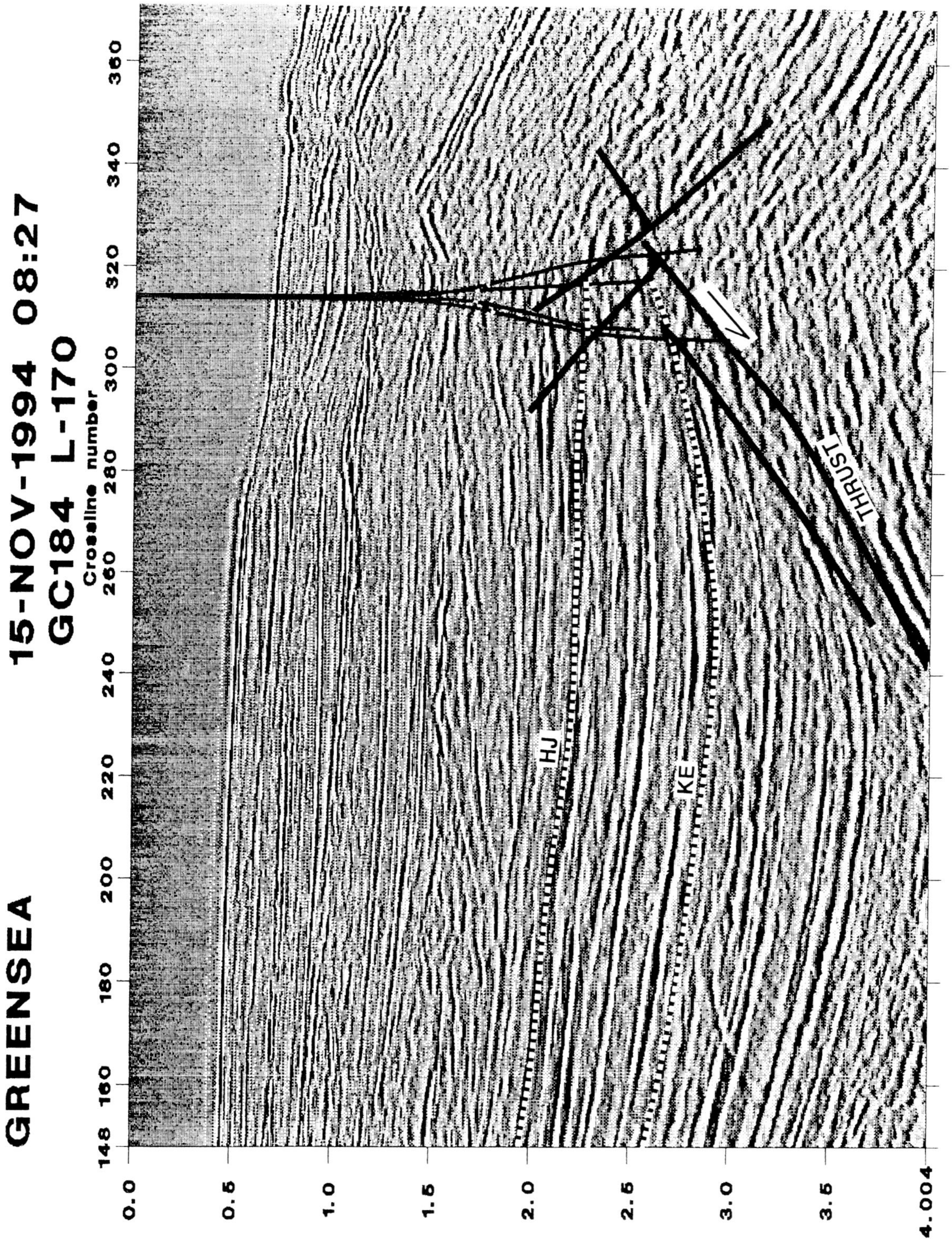

Fig. 4.—Seismic profile across Jolliet Field showing the thrust fault and position of the HJ and KE sands.

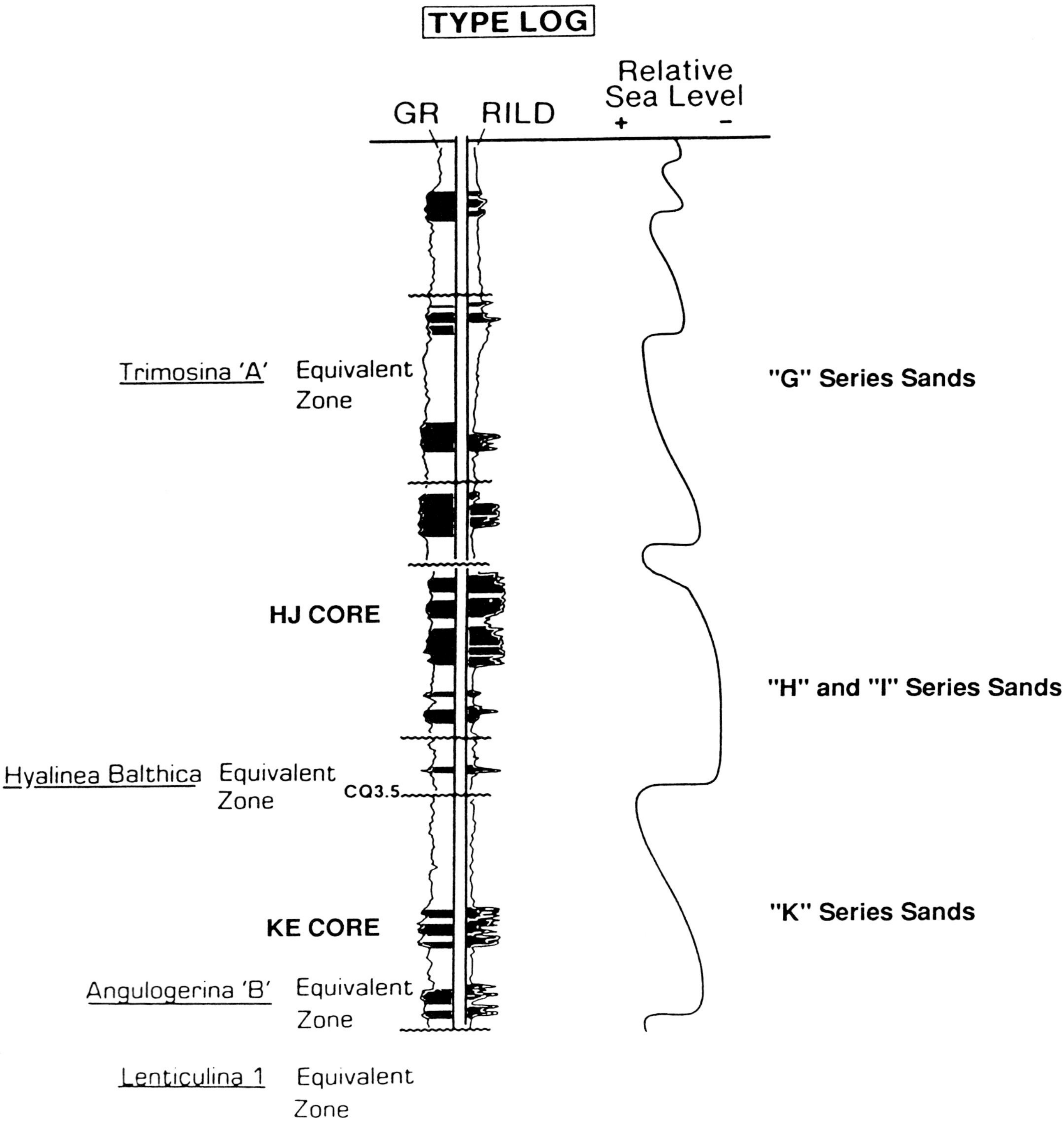

Fig. 5.—Type log showing the producing sands and the key paleontologic markers.

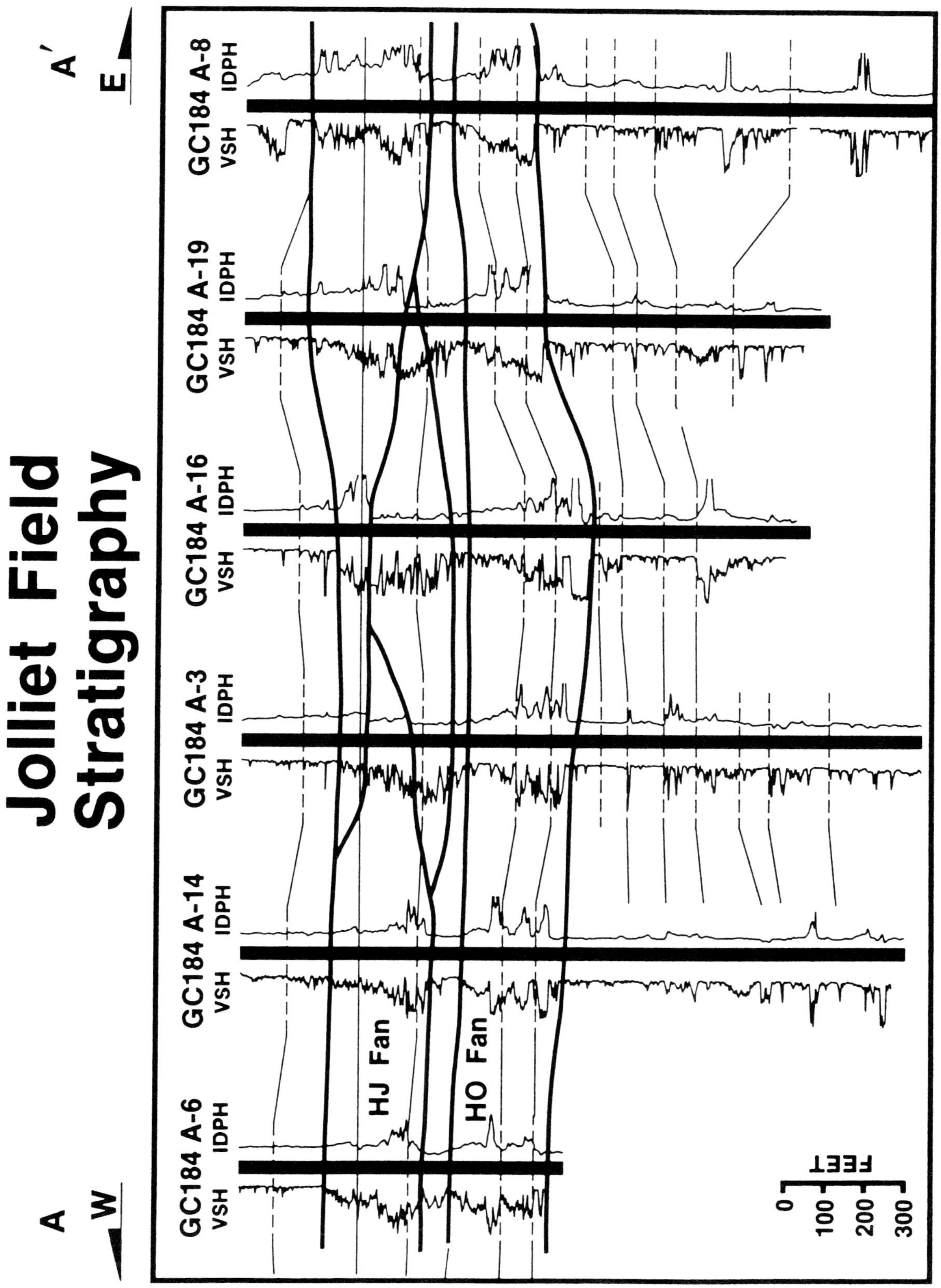

Fig. 6.—Amalgamated channel-levee deposits of the H/I series sands. Section location shown in Figure 3.

averaged 374 feet NEP and no dry holes were drilled. Possible fines production was a major concern with respect to these very fine grained reservoirs. Completions were accomplished using gravel packs and screens and these have generally been successful.

Notable technical achievements through the development phase included the predrilling program, the installation of and production from the first tension leg well platform in the Gulf of Mexico in then record setting water depth (1760 feet), and successful completion of fine grained reservoir sands.

RESERVOIR CHARACTERISTICS

Jolliet field is composed of over 104 producible reservoirs located between 6000 to 8800 feet subsea and extending over 1000 acres. The reservoirs are complexly stacked. Porosities in these sands range from 15 to 40 percent and average 27 percent. Permeability ranges from 5 to 3000 md and averages 288 md. Initial reservoir pressures above the thrust ranged from 3700 to 6820 psi, and pressures as high as 9565 psi were encountered in the subthrust section at 11,855 feet. Hydrocarbon column heights above the subthrust are generally in the 400 to 800 foot range. Column heights below the subthrust appear to be up to several thousand feet based on pressure data.

Initial production from Jolliet Field was in November, 1989, and peak oil production occurred in November, 1990 at 18,200 BOPD and 44 MMCFD. Cumulative production through December, 1993 has been 14.2 MMBO and 62 BCFG. Rates at the end of 1993 were 7100 BOPD and 34 MMCFD.

CORES

Four wells in the Green Canyon 184/228 area were cored and a total of eleven cores were taken. Special techniques were used within these relatively unconsolidated sediments in order to obtain recovery. Cores were left in PVC sleeves as they were taken from the core barrel and frozen on-site before transportation to shore. X-ray views of the sleeved cores were used for orientation prior to slabbing. Three cores from the Green Canyon 184 #7 well and one core from the Green Canyon 184 A-1 well will be discussed below.

HJ Sand in the Green Canyon 184 #7 Well

Description.—

The Green Canyon 184 #7 well was cored within the HJ interval between 6801 and 6830.5 ft measured depth. The interval comprises three distinct lithofacies (Fig. 7). The lowermost lithofacies occupies core #2 between its base and 6816.8 ft and consists of a repetition of relatively thin beds that grade upward from very fine-grained sand to silt. Some of these beds may lie in upward thinning sets (as in the intervals 6820.75-6819.2 and 6823.4-6820.75 ft). Most of the beds in this facies show a very fine, parallel lamination, which is indistinct in the coarser, sandier portions, suggesting Bouma Ta-b intervals (Fig. 8). Basal bed contacts are sharp, and a few are clearly erosional, truncating underlying strata (as at 6826.2 ft). Deformed lamina, flame structures and/or load pockets, occur at and near the contact between beds at 6828 ft. Biogenic structures are absent.

The second lithofacies occupies the interval between 6816.8 and 6811 ft, which is composed of medium- to fine-grained thin-bedded sand in which ripple-lamination predominates. The few distinctly graded sand beds in this interval are no more than a few inches thick.

The third lithofacies, which composes the uppermost interval, from 6811 to just below 6801 ft, consists of graded sand beds in the range of 1-2 feet thick that are in some places separated by thin (several inches) beds of mud. Much of the sand is ripple-bedded (Fig. 9), although both planar lamination and structureless intervals also occur. Bouma sequences appear to be common. Some of the mud beds have sharp bases (Fig. 10); others appear to grade into underlying sand. A light-colored very fine layer at 6801.75 ft may be an ash.

The three lithofacies appear to be stratigraphically intergradational. A possible break occurs at the base of the sand bed at 6816.8 ft, but this contact does not seem to be erosional and the contrast between the superjacent and subjacent strata is small.

Interpretation.—

The abundance of graded beds and Bouma sequences attest to a turbidite origin for most of the uppermost and lowermost lithofacies in the 184 #7 core. Deposition in the middle lithofacies appears to have been dominated by bed-

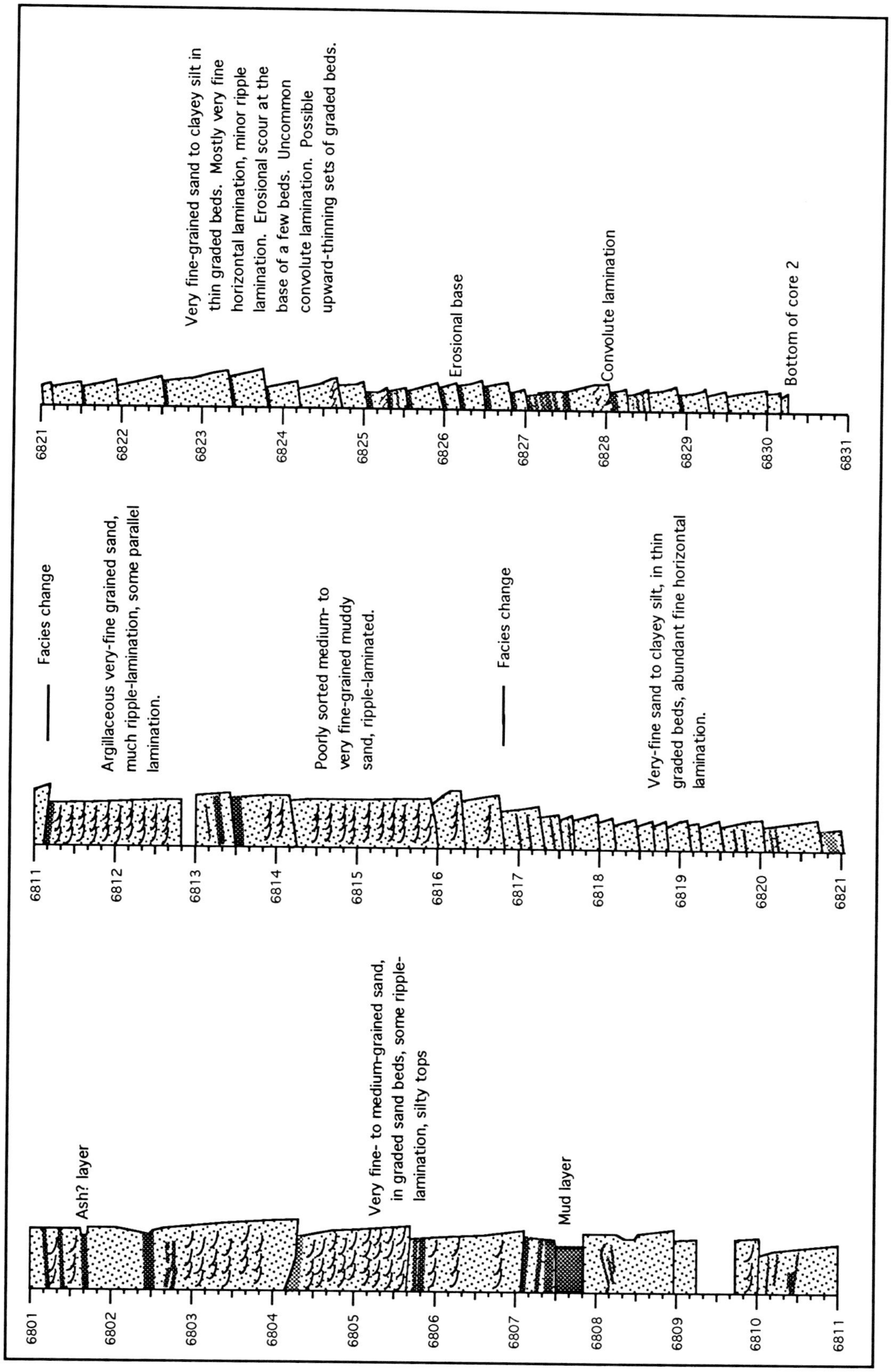

Fig. 7.—Distribution of the HJ sands in core 2, Green Canyon well 184 #7.

FIG. 8.—Photograph of core from Green Canyon well 184 #7, near 6827 ft, showing thin, finely-laminated, graded, fine-grained sand beds. Core diameter is 3 in.

FIG. 9.—Photograph of core from Green Canyon well 184 #7, near 6803 ft, showing abundant ripple-lamination. Core diameter is 3 in.

FIG. 10.—Photograph of core from Green Canyon well 184 #7, near 6808 ft, showing graded sand beds and sharp-based mud layer. Core diameter is 3 in.

load transport. The succession can be interpreted as two intervals of turbidite deposition interrupted by a period of reworking, perhaps by contour currents. An alternate possibility, however, is suggested by the difference in character of the uppermost and lowermost facies and the absence of erosional contacts. Both of these observations are consistent with an interpretation that the succession resulted from migration of a channel-levee-overbank system or the progradation of a channel/lobe system. The thin stratigraphic interval (5 feet) occupied by the middle, ripple-bedded lithofacies indicates that the bathymetric relief associated with such a system was relatively small.

Facies of similar scale in submarine canyon fill of Paleocene age exposed at Point Lobos, California (Fig. 11) (Clifton, 1981; Clifton and Hill, 1987) provides a potential analogue to the lithofacies in the core. Although mostly conglomeratic, the upper part of this fill (part of the Carmelo Formation) contains upward-fining sequences of sandstone and shale. The best exposed of these, at Weston Beach (Fig. 12) contains an upward transition (Fig. 13) from graded sandstone beds typically ten inches or more thick through about 8 feet of thin-bedded (2-6 inches thick) graded sandstone dominated by horizontal lamination and ripple bedding into mudstone with numerous thin beds of graded and commonly ripple-laminated sandstone mostly less than 2 inches thick. The thicker-bedded sandstone in the lower part of the section contains sporadic interbeds of

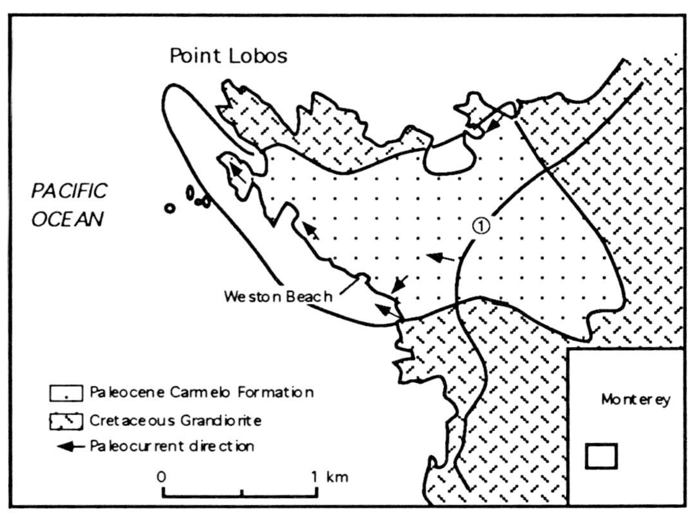

FIG. 11.—Distribution of the Carmelo Formation and Cretaceous granodiorite at Point Lobos, California, and location of Weston Beach.

FIG. 12.—Photograph of fining-up section in Paleocene submarine canyon fill at Weston Beach, Point Lobos, California.

mudstone, which like that in the top facies, is presumed to result from hemipelagic sedimentation in the intervals between flows. Bioturbation is locally intense in the finer-grained beds. Although the contacts are not erosional, the thin-bedded sandstone beds dip more steeply (to the south) the do the sub- and superjacent strata, and the orientation of ripples and ripple lamination indicates paleocurrents in this facies deviated to the south by about 30° relative to those recorded in the rock immediately above and below.

The succession at Weston Beach is inferred to reflect a laterally migrating small channel-levee-overbank system during a phase of waning deposition in the submarine canyon (Fig. 14). The thicker-bedded sandstone in the lower part of the succession is interpreted as channel floor fill. The stratal disconformity of the thin-bedded sandstone facies, a deviation in paleocurrent orientation which is consistent with a slope inclined away from the channel, and the dominance of bedload transport in this facies is consistent with a levee origin for these beds. The mudstone at the top appears to be an overbank area that received intermittent spill-over from flows in the channel.

The lithofacies in the core display a number of similarities with those described above. The thicker-bedded sandstone is inferred to be directly analogous to that in the Weston Beach succession. The intermediate interval in the core, like the inferred levee deposits at Point Lobos, was dominated by bedload transport. The facies in the lower part of the core resembles the mudstone unit at Weston Beach in its finer grain size and thinner sandstone beds relative to the other lithofacies. Like the succession at Weston Beach, the 184 #7 core section can plausibly be interpreted as the result of a small migrating channel-levee-overbank system.

The differences between the two successions can be explained in terms of grain size, slope, and rate of accumulation. Sand in the Weston Beach succession is somewhat coarser grained, which may reflect a more proximal setting, steeper slope gradients, or an initial input of coarser

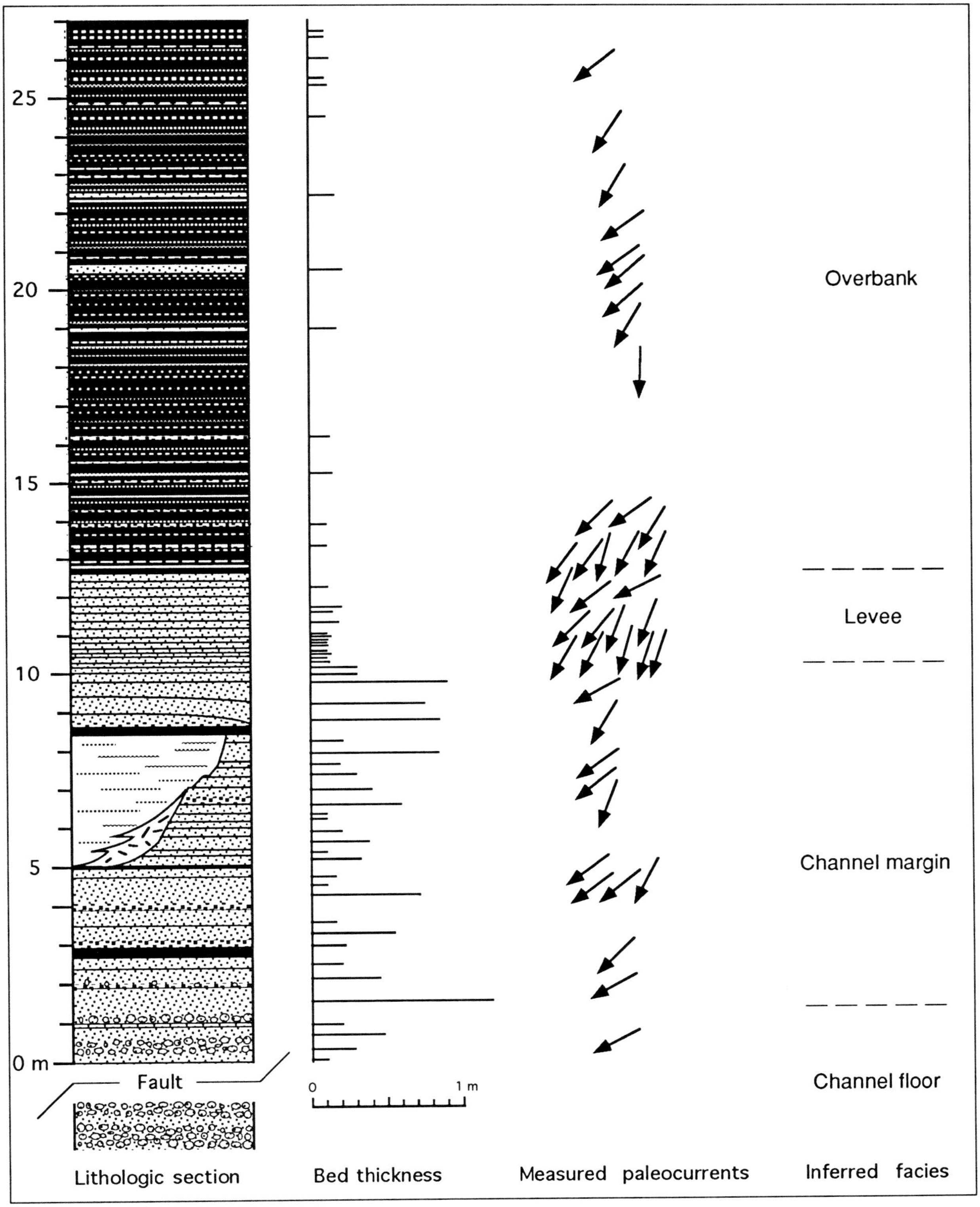

Fig. 13.—Lithologic section through fining-up succession in the Carmelo Formation at Weston Beach, Point Lobos, California. Also shown are thicknesses of all beds more than 10 cm thick, paleocurrents, and inferred depositional facies.

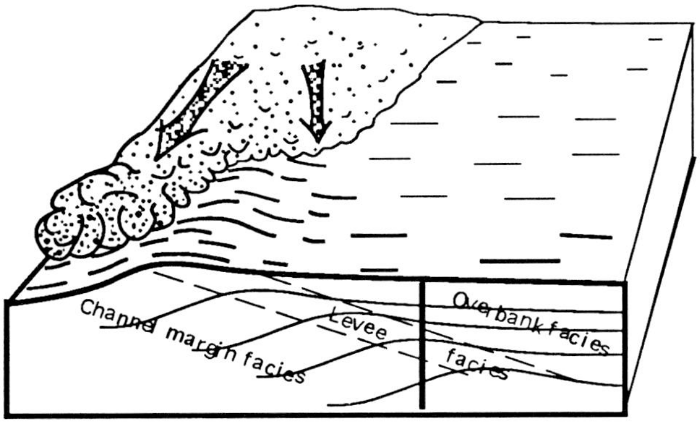

FIG. 14.—Interpretive sketch of the origin of the fining-upward succession at Weston Beach, Point Lobos, California (inferred position of section shown by heavy line). Note divergence of flow where turbidity current spills over the levee and the vertical succession produced by the migration of the channel-levee-overbank system. After Clifton (1981).

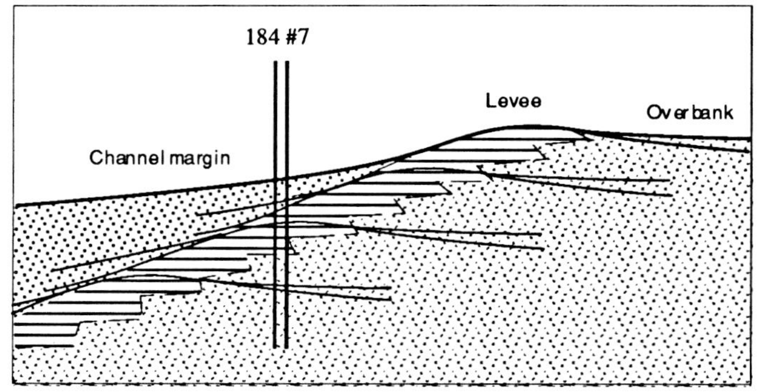

FIG. 15.—Schematic representation of a rapidly aggrading small channel-levee system producing a succession like that in the HJ sand cored in the Green Canyon 184 #7 well.

sand. The Point Lobos strata appear to have been deposited under conditions of higher energy than those in the core. The inferred levee deposits in the core are mostly ripple-bedded, whereas those in the outcrop show contain much flat bedding, suggesting flows of higher velocity. The inferred overbank deposits also suggest stronger flows in the California example, which , unlike those in the core contain much ripple-lamination. The general absence of bioturbation and hemipelagic muds in the core indicates higher rates of deposition than at Point Lobos, and the inversion of the core succession (which has the inferred channel floor facies at the top) can only be accomplished by very rapid aggradation relative to the rate of channel migration (Fig. 15).

If the HJ sand in well 184 #7 is correctly interpreted as the product of a small, rapidly aggrading levee/channel system, it is likely that it accumulated near the distal end of an intermediate slope fan (Fig. 16). Fans crossed by small distributary channel systems are known in areas of rapid sedimentation in modern base-of-slope environments and would seem likely in any mid-slope setting where the gradient decreased either within, at the end of, or proximal to a major feeder channel.

An objection to the foregoing interpretation is that known levee/overbank facies in Gulf of Mexico deep-water deposits are significantly more mud-rich than the HJ strata cored in Green Canyon 184 #7 (Shew and others, 1994). An alternative possibility is that the succession accumulated not as a channel-levee-overbank system, but as a result of progradation of a small channel across a depositional lobe. In this case, the lowermost part of the core represents deposition on an aggrading lobe, the ripple-laminated portion reflects a hydraulic jump at the channel mouth, and the uppermost part of the core, as before, is inferred to accumulated within a channel. As before, the succession in the core represents a transition from unconfined to confined flow. A prograding channel-lobe system would fit into the same depositional setting shown in Figure 16, but the requirement for extremely high rates of sedimentation is removed.

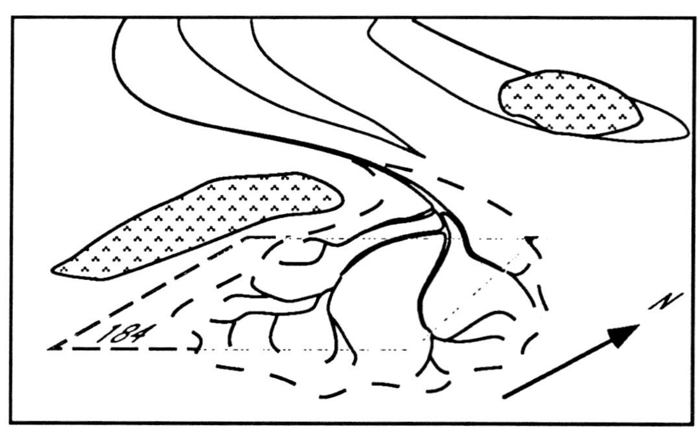

FIG. 16.—Schematic representation of the inferred environmental setting for accumulation of the HJ sand cored in the Green Canyon 184 #7 well.

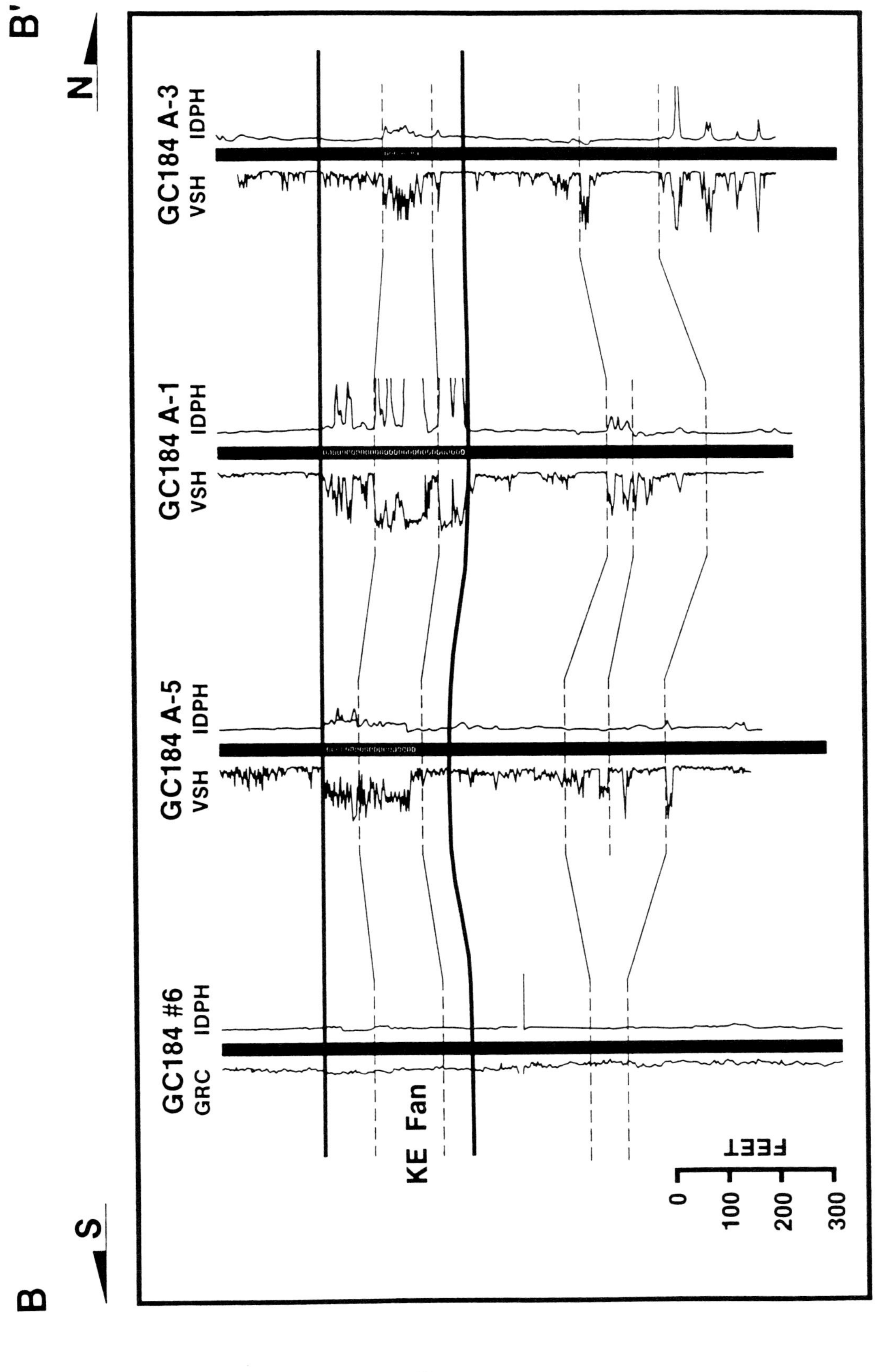

Fig. 17.—Cross-section of the KE interval. See Figure 3 for location.

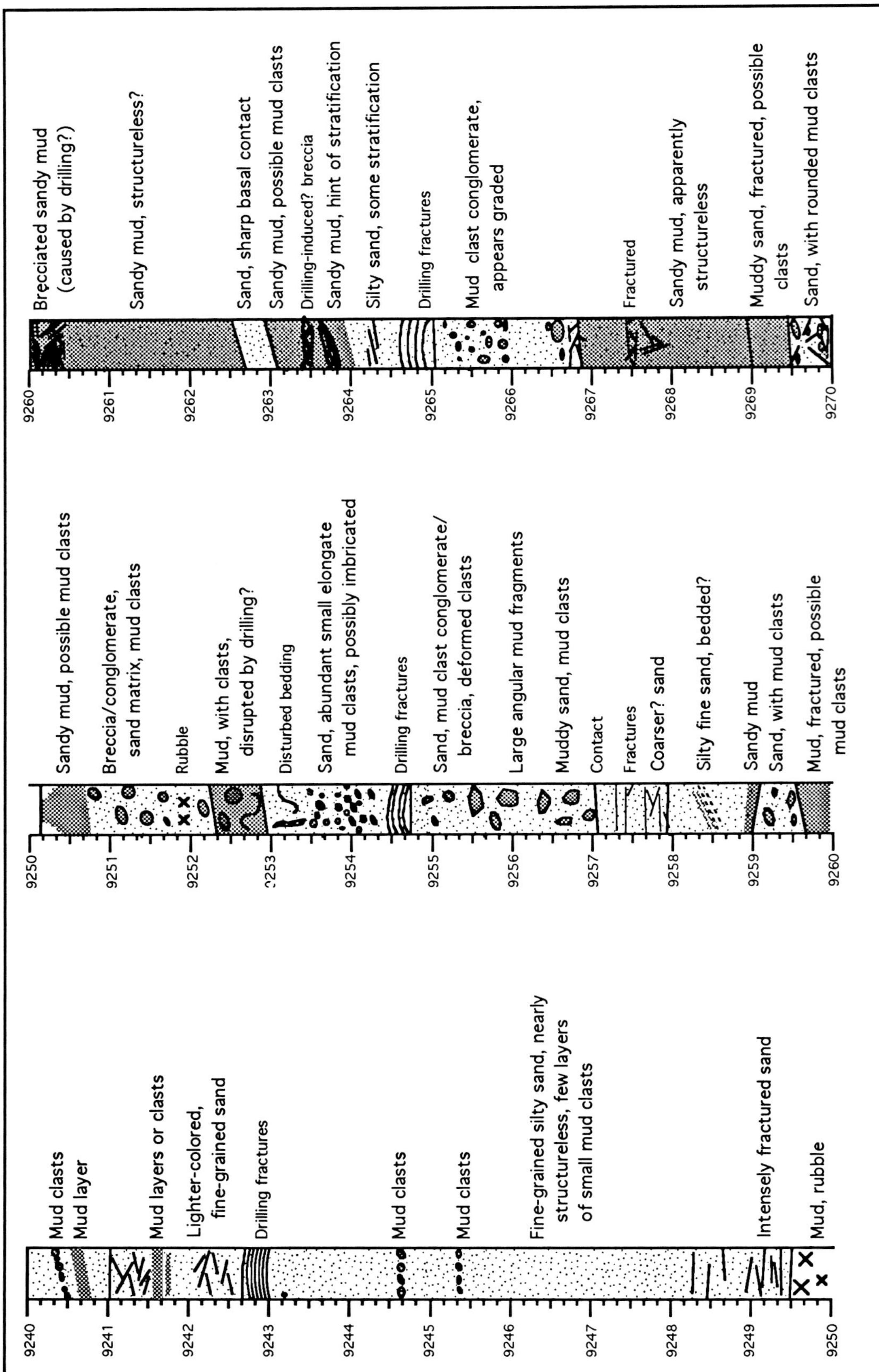

Fig. 18.—Core description of the KE-2 sand, 9240-9270 ft., Green Canyon well 184 A-1.

KE-2 Sand in the Green Canyon 184 A-1

The KE-2 sand was cored within the Green Canyon 184 #7 sidetrack well and the Green Canyon 184 A-1 well. The locations of these cores are shown in the cross section of Figure 17. The KE completion in the A-1 well has been the best completion in the field, with cumulative production of over 2.7 million bbls and 9,184 mmcfg through December 1993.

Description.—
The lower KE sand was cored in the interval 9270-9240 feet in Well 184 #A-1. Sediment in this interval consists of interbedded sand and sandy mud (Fig. 18). Mud-clast conglomerates or breccias, mostly with a sand matrix, are common.

Sandy mud, in beds as much as 3 feet thick, dominates the lower half of the core. Most of this mud shows little internal structure. Nearly all of the mud beds have sharp bases, although just below 9264 ft mud may grade into underlying sand. Some of the muddy intervals (such as 9260.6-9259.6 ft) are brecciated, probably as a consequence of breakage during drilling. Sand predominates in the upper 9 feet of core. The grain size of the sand is uniform and separate beds are difficult to differentiate. A few probably graded beds have bases at 9244.6, 9244.5, and 9242.5 ft.

In places (near 9264.8, 9254.6 and 9242.8 ft), the sand shows what appears to be a rhythmic repetition of thin beds. The orientation of the bounding surfaces, however, are perpendicular to the core in intervals where obvious stratification is at an angle, indicating that the apparent stratification probably results from drilling fracture. Other fractures that seem filled with drilling mud occur in the intervals 9250-9247.5 and 9243-9241 ft. Fine lamination is rare in the sand; some occurs in the interval 9265 and 9264 ft.

Mud-clast conglomerates or breccias are common throughout the core. Some of the may result from fracturing during drilling, but others clearly appear to be of sedimentary origin. Some, as in the interval 9267-9265 ft, consist of rounded clasts, some of which are deformed, and seem to be normally graded. Other mud-clast accumulations in the upper part of the core are thin and clearly inclined relative to the core. The interval between 9254.5-9253 ft contains abundant small mudstone clasts, many of which appear to imbricated. Oversteepened dips and synsedimentary deformation are suggested by the very steeply inclined clasts in the upper part of this interval.

Interpretation.—
The abundance of mudstone clast conglomerates, some of which appear to be deformed, combined with the massive-appearing sands and mud intervals, suggests that this interval originated in large part from debris or slurry flows. Such flows may have derived from mass failure on the margin of feeder channels or nearby slopes, as suggested in Figure 19.

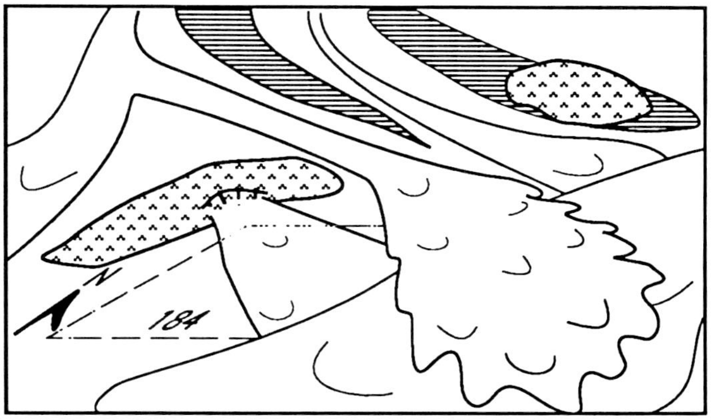

FIG. 19.—Schematic representation of the inferred environmental setting for accumulation of the KE sand cored in the Green Canyon 184 A-1 well. Slumps may have been derived from nearby salt intrusions, from feeder channel walls , or from sources higher on the slope.

KE-2 Sand in the Green Canyon 184 #7

Description and interpretation.—

The KE Sand was cored in two intervals, 8536-8526 ft and 8489.7-8485 ft, in the Green Canyon 184 #7 sidetrack well. Much of the sediment in these cores is highly disturbed (Fig. 20), but much of the disruption may be due to the drilling process. In general, the sands in these cores are better stratified than those in the 184 A-1 well and may include a few Bouma sequences. Deposition of the KE sands in the sidetrack well may have been more influenced by turbidity currents than by debris flows.

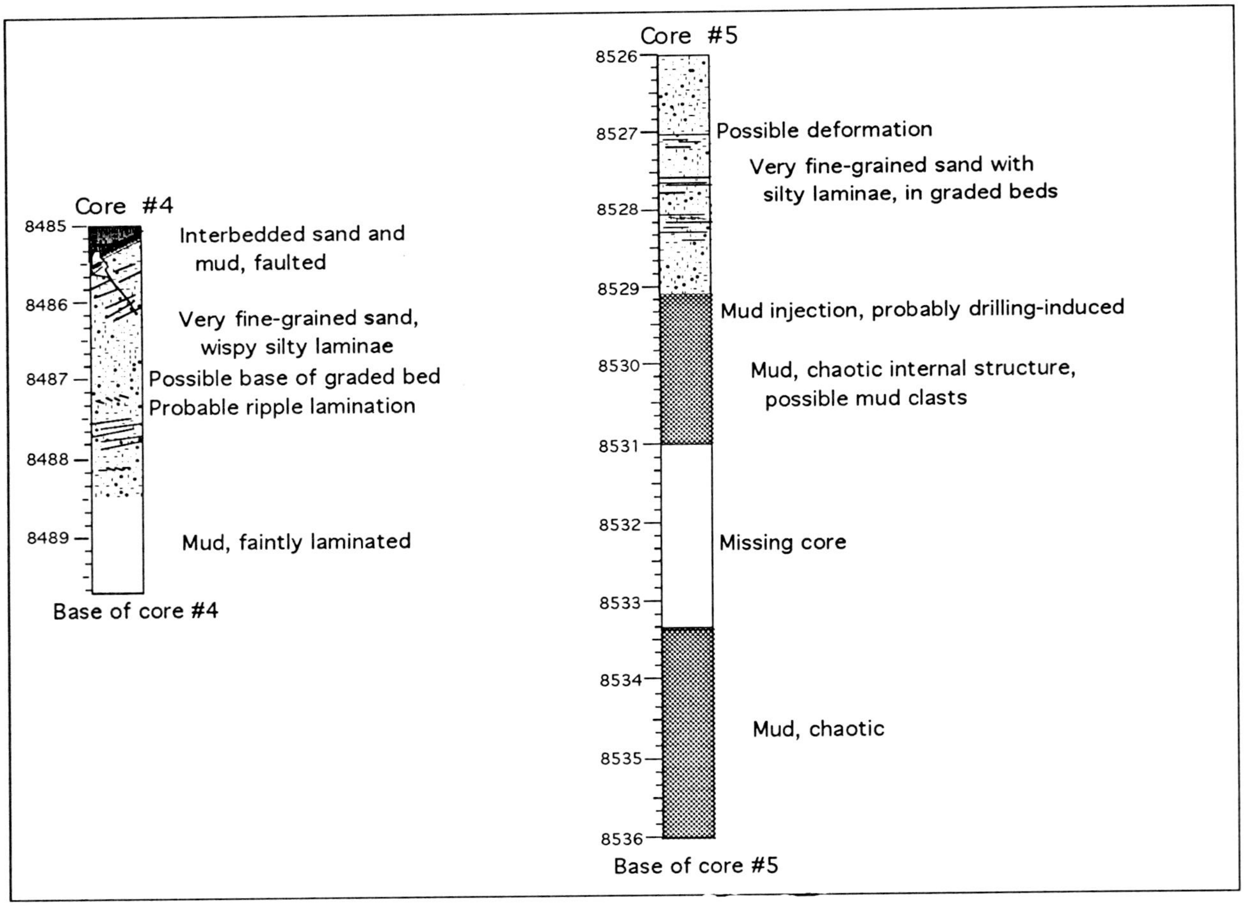

FIG. 20.—Description of the KE sands in Green Canyon 184 #7 sidetrack well.

APPRECIATION

Appreciation is expressed to management at Conoco, Texaco, and Oxy for allowing this material to be published and presented. Core Labs is thanked for their help in preparing some of the core photos. Karen Christensen helped with the paleogeographic reconstruction of the Green Canyon area.

REFERENCES

CLIFTON, H. E., 1981, Submarine canyon deposits, Point Lobos, California: *in* Frizzell, V., ed., Upper Cretaceous and Paleocene turbidites, central California coast; Pacific Section SEPM Field Trip Guidebook, Field Trip Number 6, San Francisco Annual Meeting p. 79-92.

CLIFTON, H. E., and HILL, G. W., 1981, Paleocene submarine canyon fill, Point Lobos, California: Geological Society of America Field Guide, Cordilleran Region, p. 239-244.

COOK, D., and D'ONFRO, P., 1991, Jolliet Field thrust fault structure and stratigraphy, Green Canyon 184, offshore Louisiana: Transactions Gulf Coast Association of Geological Societies, v. 41, p. 100-121.

HELLER, P. L., and DICKINSON, W. R., 1985, Submarine ramp facies model for delta-fed, sand-rich turbidite systems: American Association of Petroleum Geologists Bulletin, v. 69, p. 960-976.

NUNN, J. A., and SASSEN, R., 1986, The framework of hydrocarbon generation and migration, Gulf of Mexico continental slope: Transactions Gulf Coast Association of Geological Societies, v. 36, p. 257-262.

ROBERTS, H. H., COOK, D. J., and SHEEDLO, M. K., 1992, Hydrocarbon seeps of the Louisiana continental slope: seismic amplitude signature and sea floor response: Transactions Gulf Coast Association of Geological Societies, v. 42, p. 349-361.

SASSEN, R., BROOKS, J. M., MACDONALD, I. R., KENNICUTT, II, M. C., GUINASSO, JR., N. L., and REQUEJO, A. G., 1993, Association of oil seeps and chemosynthetic communities with oil discoveries, upper slope, Gulf of Mexico: Transactions Gulf Coast Association of Geological Societies, v. 43, p. 349-355.

SHEW, R. D., ROLLINS, D.R. TILLER, G. M., HACKBARTH, C. J., AND WHITE, C. D., 1994, Characterization and modeling of thin-bedded turbidite deposits from the Gulf of Mexico using detailed subsurface and analog data: *in* WEIMER, P., BOUMA, A. H., and PERKINS, B. F., eds., Submarine fans and turbidite systems, sequence stratigraphy, reservoir architecture, and Economic Paleontologists and Mineralogists Foundation, 15th Annual Research Conference, Houston, TX, p. 327-334.

RESERVOIR CHARACTERIZATION OF THE "S" SAND AUGER FIELD, GARDEN BANKS 426, 427, 470, AND 471

P. W. BILINSKI[1], D. T. MCGEE[1], D. S. PFEIFFER[1], AND R. D. SHEW[2]
[1]*Shell Offshore Inc., New Orleans, Louisiana*
[2]*Shell Oil Company, Houston, Texas*

ABSTRACT

Auger Field is a 1987 deepwater discovery located 214 miles (344 km) southwest of New Orleans, Louisiana. The field consists of five productive horizons with the lowermost "S" sand reservoir containing an estimated 120 MMBOE, which is over half of the total field reserves. The Miocene "S" sand is oil productive from layered and amalgamated sheet sands that were deposited in the Auger minibasin. Onlap onto a paleoridge and faulting are the trapping mechanisms. The sands thicken into the basin away from the onlap as evidenced by both well and seismic data. The seismic data further indicate that the sands extend below the oil/water contact and should provide some aquifer support for pressure maintenance. The accumulation has a 1200 ft (366 m) hydrocarbon column based on an estimated oil-water contact from seismic at -19,650 ft (5990 m). The "S" sand is mostly a slightly consolidated, lower fine-grained sand with average porosity and permeability of 25 percent and 150 millidarcies, respectively. Production began in 1994 from a tension leg platform located in 2,862 ft (872 m) of water. High individual well rates have been established from these laterally continuous sands with average rates of 10 MBOPD and 20 MMCFGPD.

INTRODUCTION

Auger Field is located 214 miles (344 km) southwest of New Orleans. It includes the Garden Banks OCS Lease Blocks 426, 427, 470, and 471 (Fig. 1). Auger was discovered in 1987 with a vertical discovery well and two sidetracks designed to test multiple, potential reservoir levels. Stacked pays with volatile oil and retrograde gas, as defined by McCain and Bridges (1994), were penetrated in five intervals ranging from -15,500 ft (4725 m) to greater than -19,000 ft (>5790 m); see Figure 2 for a type log.

A 3D seismic survey was shot over the discovery in late 1987. This survey was shot with a 12,000 ft (3658 m) seismic cable, which was the industry standard at the time. In 1988, a follow-up appraisal well with one sidetrack was drilled in Block GB 426, and proved up the areal extent and the quality of reservoirs in that block. A second, long-cable (18,000 ft; 5486 m) 3D survey was deemed necessary to adequately image the deeper reservoirs ("Q," "R," and "S"). This 3D survey was shot and processed in 1990 and dramatically improved imaging of these reservoirs.

Beginning in 1990, during fabrication of the Tension Leg Platform (TLP), ten predrill wells were drilled by the SONAT George Richardson. Seven of the wells targeted the "S" sand and one well each was drilled for the "N1," "O," and "R" sands. Each well was temporarily abandoned after the deepest casing was run. The wells have been re-entered with the TLP's drilling rig and completed in their target reservoirs (Bilinski et al., 1992). Completion operations will continue through November of 1994. First production started April 15, 1994 from the A-6 well completed in the "S" sand. The Auger TLP set a world-record water depth, being placed in 2862 ft (872 m) of water.

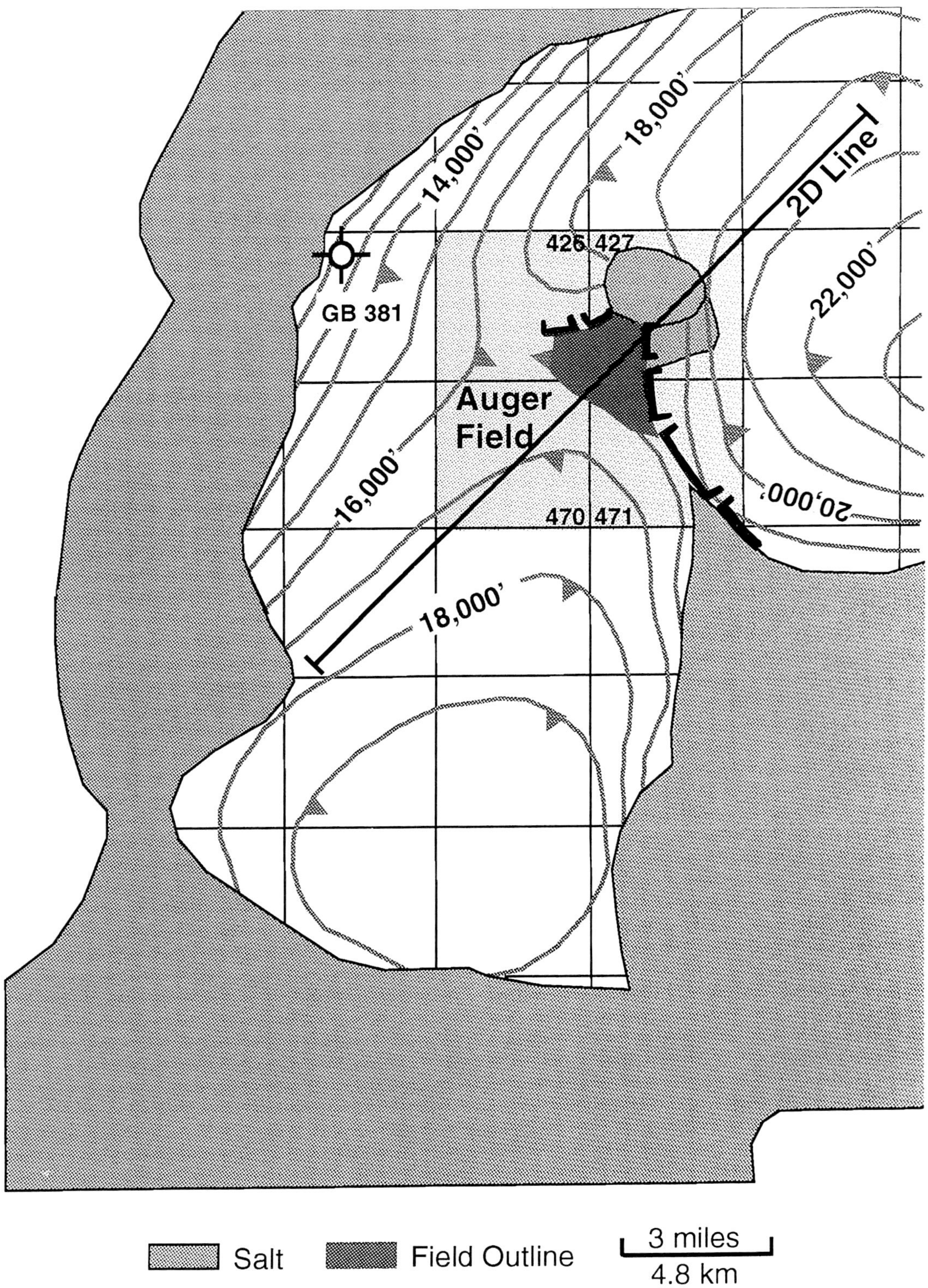

FIG. 1.—Structure map of the shallow "O" Sand level in the Auger Basin area illustrates the minibasin shape and location of Auger Field within the basin. The location of the 2D seismic line, which is the basis of the schematic drawing in Figure 3, is also shown.

AUGER BASIN SETTING

Auger Field is located in the northwest quadrant of Auger Basin, which is one of a series of connected minibasins (Fig. 1) in the Garden Banks area. The field's trapping elements include onlap onto a deep paleoridge, a down-to-the-east fault system, and a down-to-the-north fault system. A shallow piercement salt dome that is slightly overhung to the south is most likely related to the deeper paleoridge, but is not a primary trapping element for the pay section in Auger Field.

The basin is almost completely surrounded by shallow tabular salt bodies that have flowed together into nearly continuous salt masses (Fig. 1). The timing of the emplacement of tabular salt post-dates the deposition of the reservoir sands in the field and is late in the history of the basin.

Salt tectonics prior to the tabular phase of salt deformation are related to the field's formation. The structural level depicted in Figure 1 is near the "O" reservoir level, one of the shallower reservoirs in the field. The map shows the asymmetrical nature of the basin with dip toward the south. Throughout the Pliocene, active salt withdrawal and sediment ponding occurred at the southern end of the basin. Auger Basin's structural apex, a paleohigh, was located in the northwestern quadrant of the basin. Today, this area is at least 4,000 ft (1220 m) structurally higher than Auger Field. The paleoridge adjacent to Auger Field formed an onlap margin along the eastern flank of the basin (Fig. 3).

A schematic of a 2D seismic line (Fig. 3) crossing Auger Basin illustrates the history of the basin-fill and how it changed through time. On the schematic, the field pay section is located between the "N1" level and the "S" levels in the expanding section near the paleoridge. Above the field's pay section, the basin geometry changes and interval thicknesses are nearly constant (Fig. 3). The locus of expansion due to sediment deposition and active salt withdrawal shifted away from the Auger Basin during deposition of these younger sediments.

FIELD OVERVIEW

Auger Field consists of five major, stacked pay intervals of Pliocene and Miocene age. Depths range from greater than -20,000 ft (>6096 m) to -15,000 ft (4572 m) (Fig. 2 and 3). The deeper pay sands contain volatile oils with API gravities of 34 to 38°. The shallower pay sands contain retrograde gases. The "S," "R," and "Q" reservoirs are dominated by sheet sands. Above the "Q" sand, the reservoir architectures shift from dominantly sheet geometries to amalgamated channel geometries for the "O" and "N1" reservoirs. The terms used in this paper for turbidite depositional geometries follow those discussed by Steffens (1993). Sheet sands consist of both amalgamated and layered sands with the difference in the vertical connectivity. Log motifs of coarsening and thickening-upward sand packages in the lower sands change to thinning and fining-upward packages in the "O" and "N1" sands. The change in depositional styles within the field reflects the progradational nature of the basin-fill and a shift from active lobe-dominated sedimentation in a rapidly subsiding basin to channelized sedimentation in a slowly subsiding basin, with minimal accommodation space (Mutti and Ricci Lucci, 1972; Mutti and Normark, 1987; Normark, 1978; Walker, 1978).

Reservoir quality in all five pay sands is very good, but porosity and permeability decrease with depth, which is primarily a result of increased compaction at greater depths.

"S" SAND RESERVOIR

General Description

The "S" reservoir has a gross thickness greater than 200 ft (61 m) with an average net/gross of 0.8. The "S" sand consists of multiple coarsening and thickening-upward packages of sheet sands that may be separated into two members: the lower S2 and the upper S1. Both of these members contain volatile oil. The hydrocarbons are trapped against the upthrown side of the down-to-the-north fault and pinch out to the northeast near and along the paleoridge (Fig. 4). The accumulation has a hydrocarbon column height of 1,200 ft (366 m). Whole core was taken in Well 470 No. 1 and in the second pre-drill well, No. A-2. These cored wells and well logs, in addition to the seismic data, are used to interpret the external and internal geometry of the "S" sand.

Reservoir Geometry

The "S" reservoir can be internally correlated over the entire extent of the reservoir. It may be separated into the lower S2 and the overlying S1 members by a thick, continuous mudstone. The well control is located mostly along the depositional dip (structural strike) of the field (Fig. 5); the sediment source and primary depositional flow direction are interpreted to be from northwest to southeast as observed by thickening and thinning trends on seismic and from well logs. The high degree of lateral

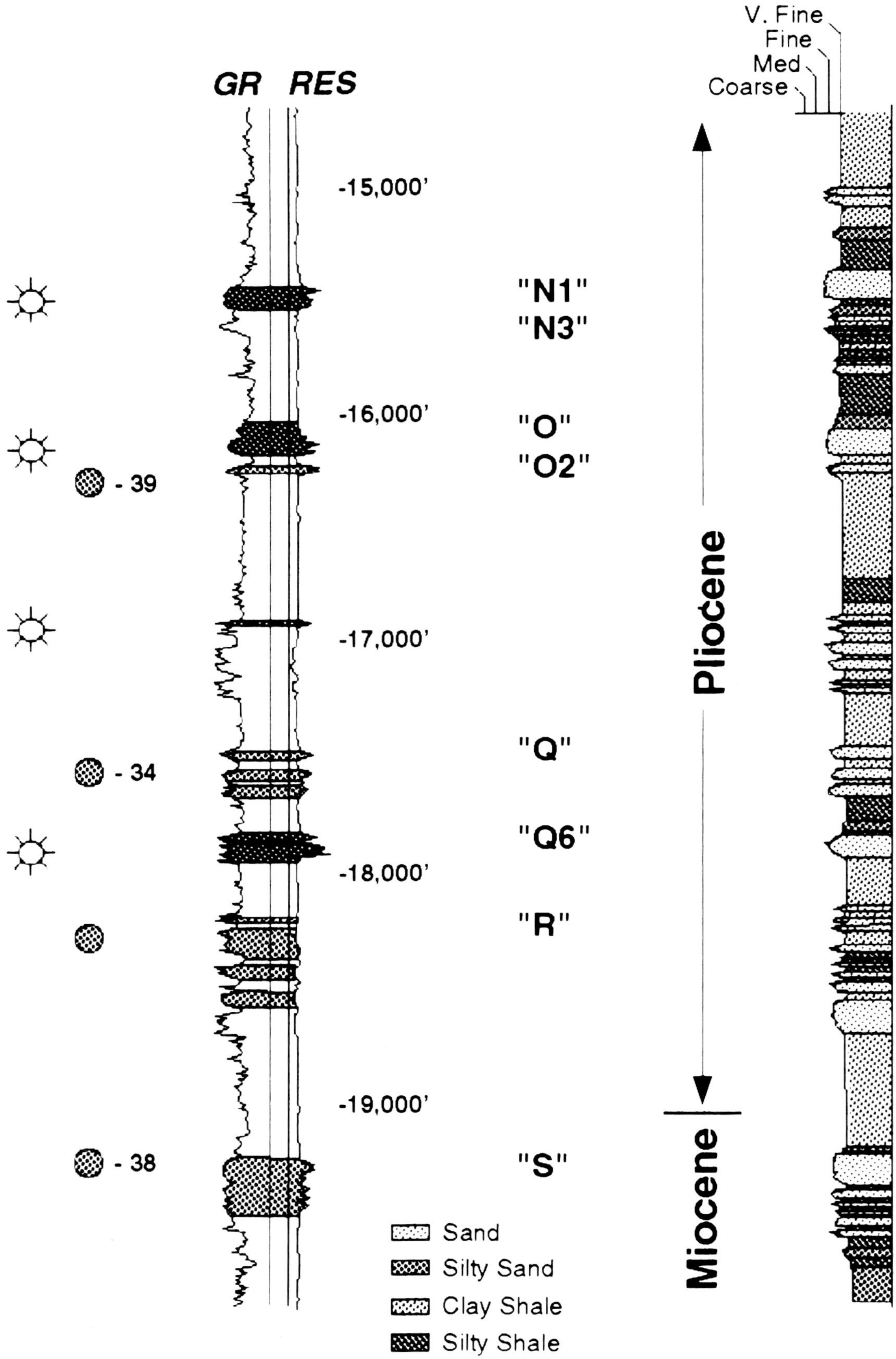

FIG. 2.—Type log of the pay intervals penetrated in Auger Field. Five major horizons (N, O, R, Q, and S) are hydrocarbon-bearing. Oil and gas are both present; the "S" sand contains greater than half of the total reserves.

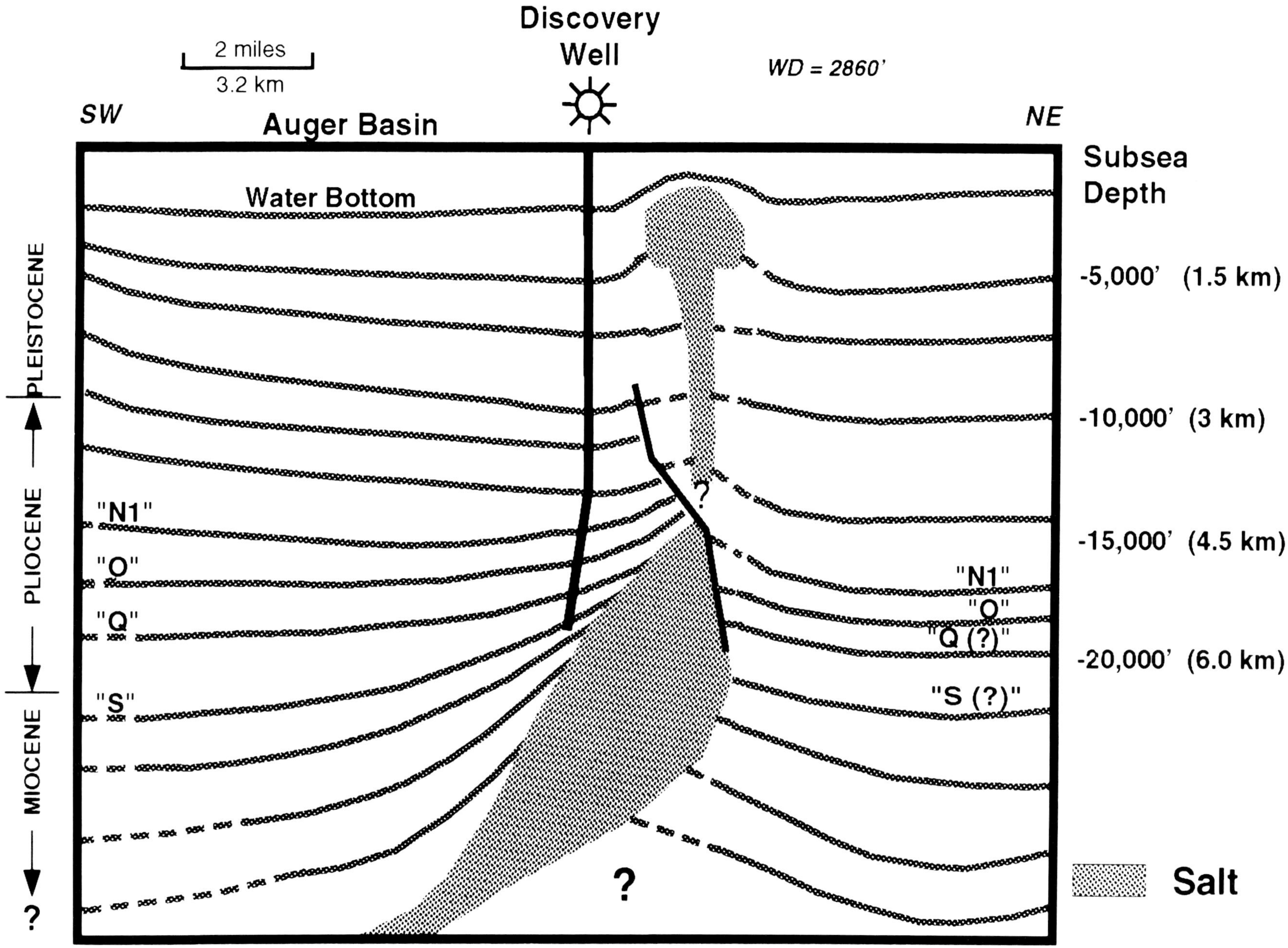

FIG. 3.—Schematic drawing of 2D seismic line (in time) crossing the Auger Basin from NE to SW (see Figure 1 for location). The depth conversion shown on the right axis used well checkshot velocities. The "S" sand is the deepest of the hydrocarbon productive zones in Auger Field.

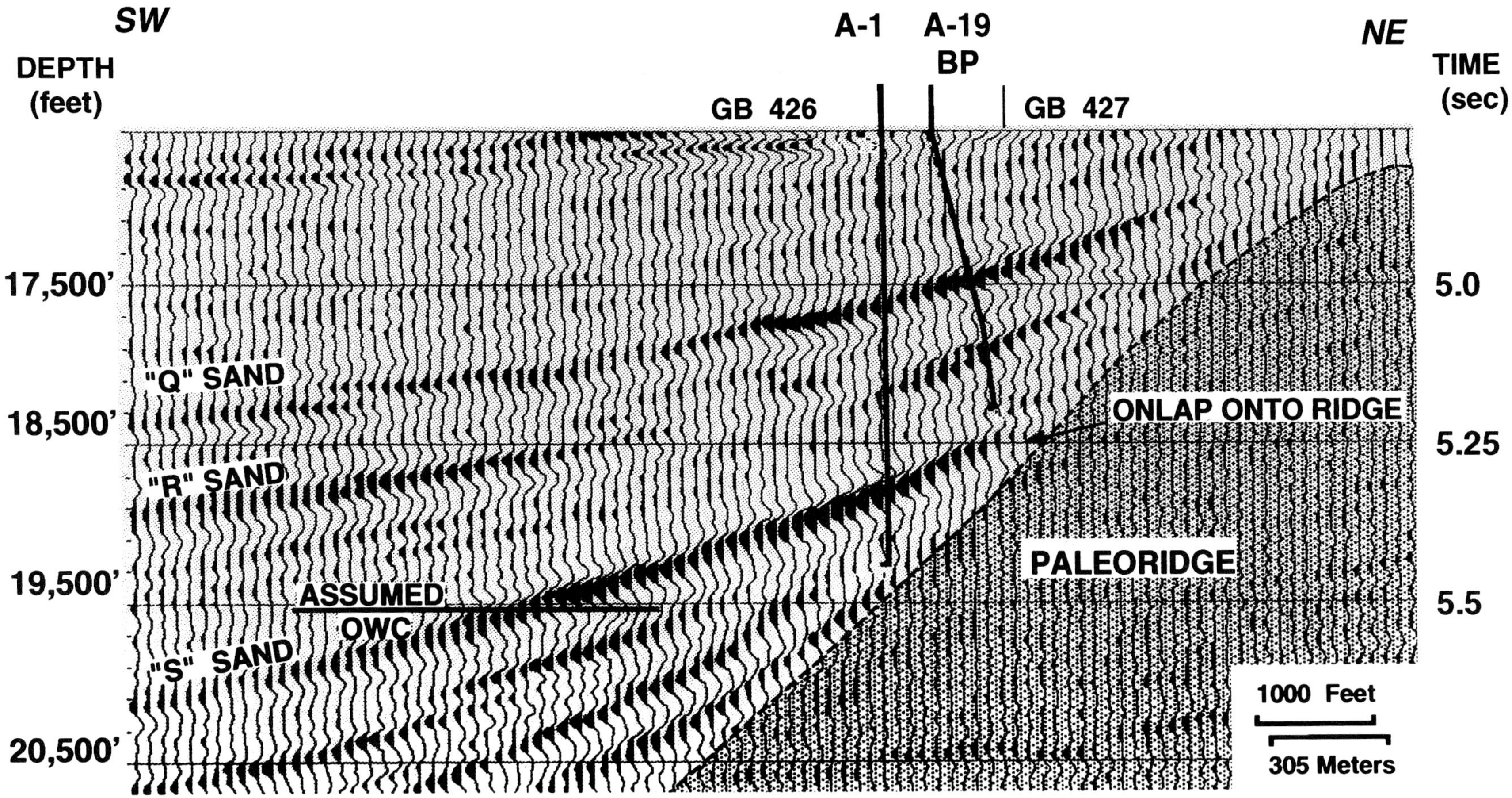

FIG. 4.—Structural dip-oriented seismic line from the 3D survey covering the "S," "R," and "Q" reservoirs. The line is along structural dip (NE to SW); the location of the dip line is shown in Figure 5.

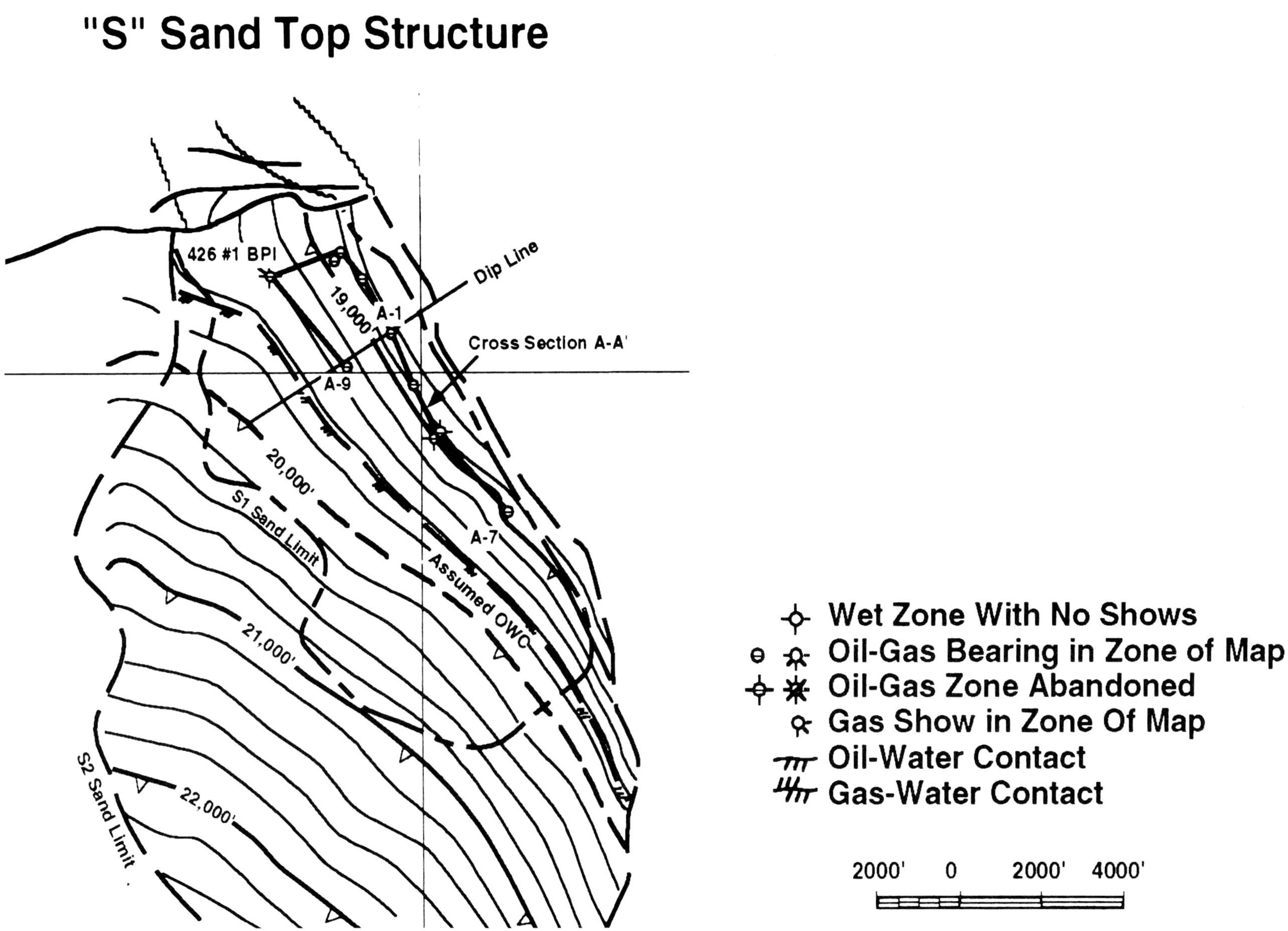

FIG. 5.—Top structure map of the "S" Sand. The map was constructed using seismic data, which was depth converted using well checkshot velocities and fit to well control. Two lines of section are shown on the map: the section of Figure 6 is along structural dip, while the cross section of Figure 7 is along structural strike but depositional dip. The sediment source direction is from the northwest.

continuity, illustrated in Figures 6 and 7, is a product of the actively subsiding basin providing sufficient accommodation space for the incoming/prograding lobe sediments. Initial sediments were deposited as lower energy, continuous layered sheets. Increasing sand content and deposition of amalgamated sheets occur upward in the section as progradation of the turbidite system continued. Figure 8 illustrates the interpreted depositional model for the "S" sand with the change from the lower layered sheet sands to the amalgamated sands. More quiescent periods of deposition, which are represented by the thin bed deposits separating the sandier section, occur throughout the interval and can also be correlated across the basin. One thicker, highly correlative mudstone with minor thin beds occurs in the section and is used to subdivide the "S" sand into the S1 and S2 members. There are local areas of thinning of the "S" sand both at the bottom and top of the sand (Fig. 6 and 7). In Well No. A-7, thinning at the base of the sand is related to onlap onto a carbonate deposit formed in response to the presence of a paleoseep (Shew et al., 1993). Thinning at the top may be related to post-depositional scour and/or to thinning on the paleohigh.

The "S2" member contains both massive and thin-bedded (1-3 ft thick; 0.3-0.9 m) intervals of upper very fine to lower fine-grained subarkosic to arkosic sands (Fig. 9 and 10). The thin beds are concentrated in the center of the member (Fig. 6, 7, and 9). The thin beds consist of partial Bouma (1962) sequences, including a few massive (Ta) subdivisions but with more occurrences of parallel-laminated (Tb), rippled (Tc) and/or convoluted sedimentary structures. The sedimentary structures are often disturbed by soft sediment deformation. Floating shale rip-up clasts occur near the top of the sequences and are interpreted to be intraformational and related to scouring during the same depositional event (Mutti and Nilsen, 1981). These thin beds may represent a time of less robust sheet flows or more distal portions of flows that were centered in the basin axis to the west. The top and basal portions of the "S2" member are massively bedded and of similar quality to the "S1" member. These beds probably represent deposition in the main axis of turbidite flow. The relationship between the various flows demonstrated by the sequence of massive, thin-bedded, and then massive beds probably was compensational in the style described by Mutti and Sonnino (1981). The deposition of the first massive sand deposits formed a local paleohigh which diverted the subsequent deposits laterally. The final massive deposits filled in remaining paleolows.

The "S1" member is separated from the "S2" member by a 15 to 20 ft (4.6 to 6.1 m) thick mudstone layer that is correlative across the field (Fig. 6 and 7). In core (Fig. 10), this layer contains 0.5 to 4 inch (1.3-10 cm) partial Bouma (1962) sequences, mostly rippled to parallel-laminated silt and mud (Td), encased in turbiditic muds. These muds are silt-rich, contain few or no microfauna, and contain small graded beds. This mudstone layer is interpreted to be a vertical permeability barrier throughout the reservoir. The "S1" contains massive beds (Fig. 10 and 11) of lower fine-grained, subarkosic to arkosic sands with few to no visible sedimentary structures. Whole core from the A-2 well contains almost 30 ft (9 m) of massive "S1" sand with no visible sedimentary structures, dominantly massive (Ta) packages. Rare amalgamation surfaces in these sheet sands, visible in the GB 470 #1 core, are distinguished by minor grain size changes and provide some indication of bedding sequences. High energy deposition has led to mostly sand on sand contacts with little change in permeability across the contacts. There is likely local scour in the sands. Figure 11 illustrates the log response of the massive upper "S1" member and the grading downward into slightly finer sand with increase in clay content.

Although the core and logs provide detailed correlations and internal geometries of the "S" sand, seismic data are extremely important in imaging the distribution of the sand laterally. As mentioned, the wells are located along depositional dip adjacent to and along the updip margin of the onlap. Well placement was based on locating the wells updip using seismic. The seismic event of the "S" sand is moderately continuous onlapping abruptly onto the paleoridge. The distance from the A-1 well to the termination of the seismic event is about 1,000 ft (305 m), and over this interval, the sand thickness decreases from more than 200 gross ft (61 m) to zero feet (Fig. 7). It is interpreted that the "S1" member offlaps the paleoridge (Fig. 7). Equally important to the updip sand event termination is the downdip imaging of the sand to determine the oil/water contact as well as the possible occurrence and size of an aquifer for reservoir pressure support. Wet sand or the oil-water contact has not been penetrated to date. The assumed water level of -19,650 ft (5990 m) is visible based on amplitude and wavelet character changes on the seismic (Fig. 4). The lowest known hydrocarbons are at -19,614 ft (5978 m) in the GB 426 #1 BP #1 well, the most northwesterly well. Detailed mapping of the seismic event indicates that the "S2" probably has the most extensive aquifer. The "S1" aquifer may be smaller, covering about half of the

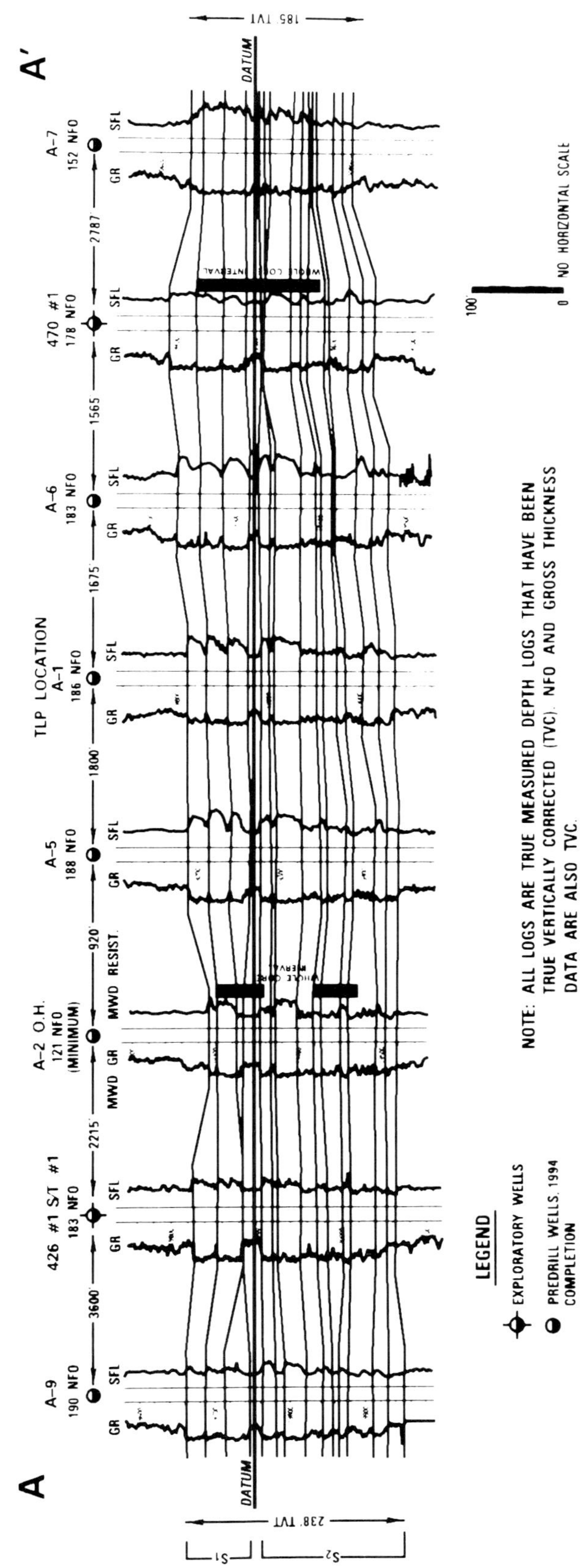

FIG. 6.—True vertical thickness cross section of the "S" sand along depositional dip (structural strike). The section is hung on the mudrock interval separating the "S1" and "S2" members. Well locations are shown in Figure 5. Note the lateral continuity of the S1 and S2 members. There are areas of thinning, in particular in the A-2 OH and in the A-7.

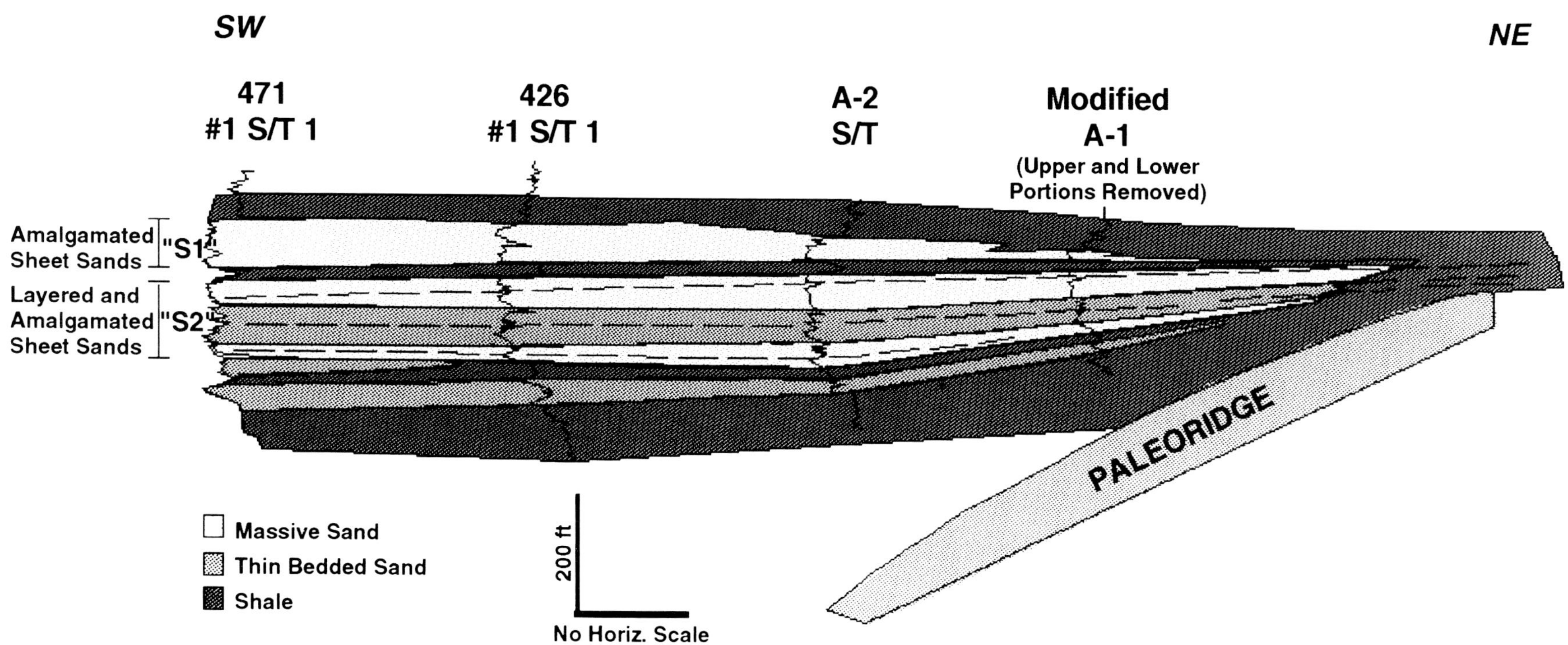

FIG. 7.—"S" sand schematic depositional strike (structural dip) cross section. Wells shown in section are located based upon interpreted position in depositional model (Figure 8) rather than true spatial position. Note the high degree of continuity of the sands with some onlap onto the ridge and the continuous low energy unit separating the lower S2 from the S1 sand.

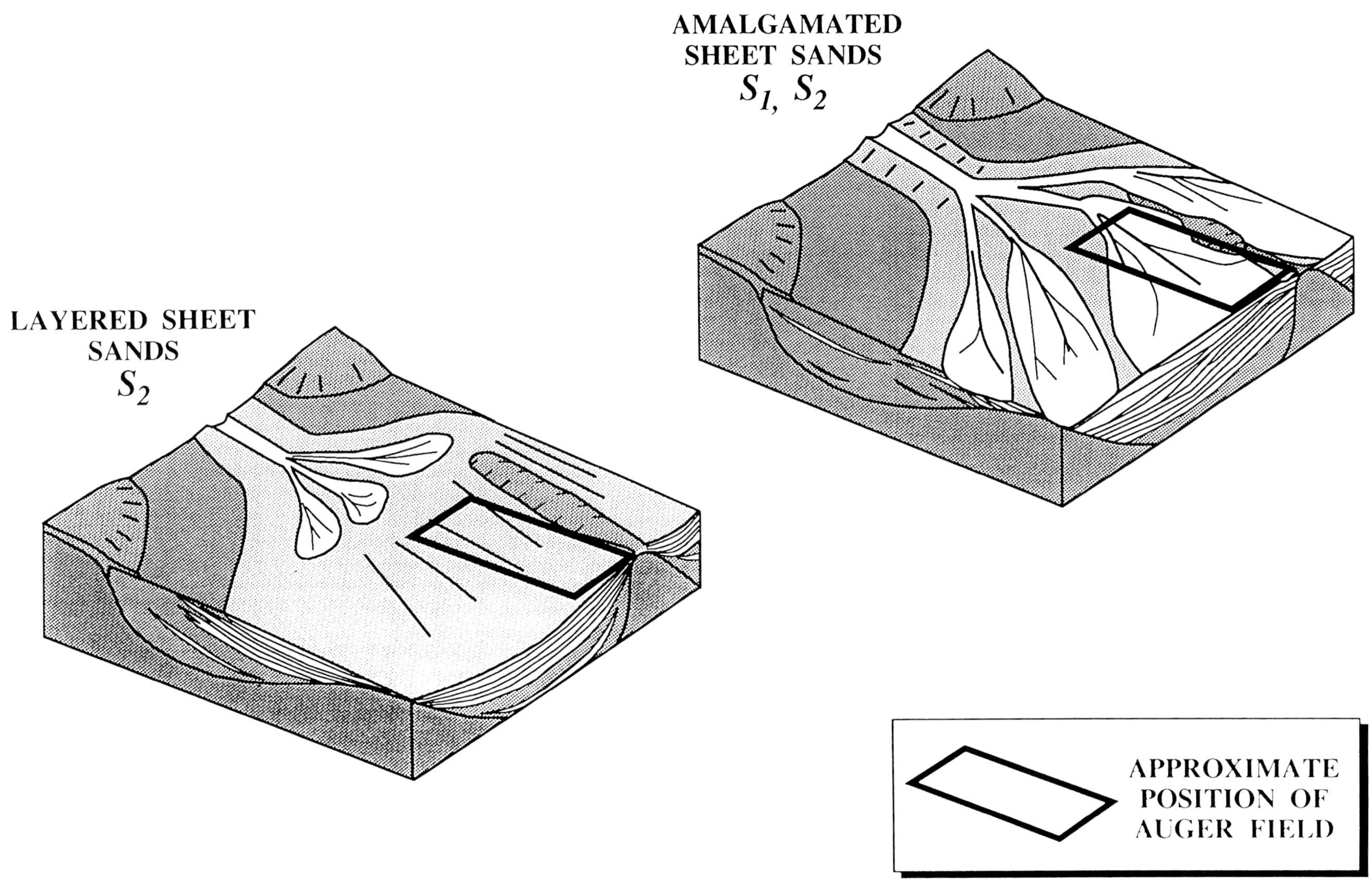

FIG. 8.—Depositional model for "S" reservoir. Individual members are listed next to their interpreted depositional geometries. These are dominantly layered and amalgamated sheet sands. Layered sheet sands were deposited first followed by more amalgamated sheet sands. Thin beds are present representing periods of decreased turbidite flows.

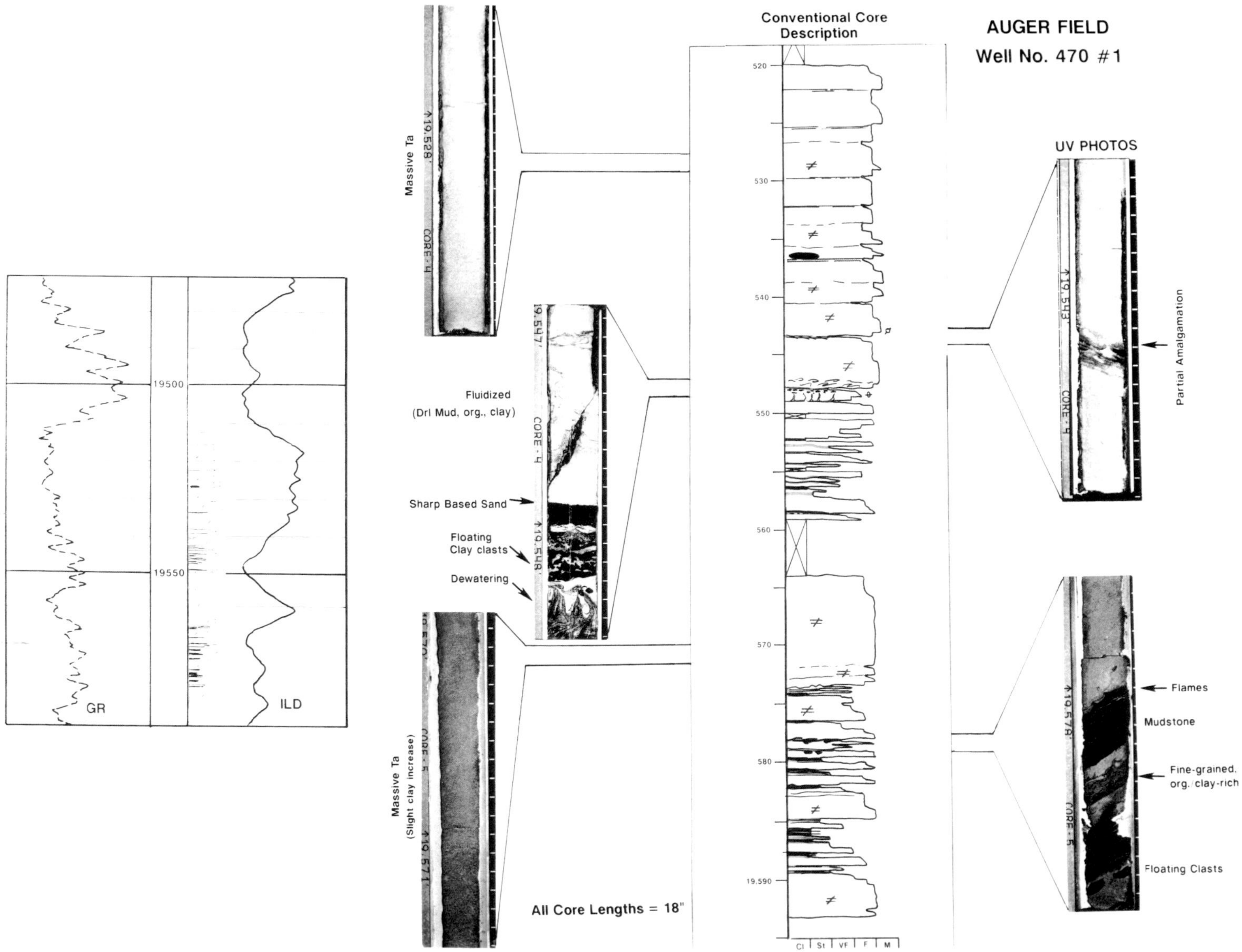

FIG. 9.—Logs and core illustrate the presence of lower energy, thinly bedded sands and silts with increased mudstone that separate the more massive amalgamated sands. Coarsening upward trends are obvious locally within the core, particularly from 19,510 ft (5946 m) to 19,553 ft (5960 m). The massive Ta, amalgamated sands can be subdivided into layers with detailed core descriptions but the net-to-gross is close to 1.0 and there are no significant vertical barriers. The thinner bedded zones have floating clasts, dewatering features, flames, graded beds, and a lower net-to-gross with poorer reservoir quality.

FIG. 10.—Three of the massive to more thinly bedded cored intervals illustrate the massive, high quality amalgamated sand and the lower quality thin-bedded and laminated, clay-rich sands. UV light photographs were used for these photos. The dominantly mudstone interval that separates the "S1" and "S2" members is in part shown in the core at 19,465 ft. The laminae are mostly rippled, graded and terminate in turbiditic mudstones. Dewatering features and intraformational clasts occur below 19,548 ft and flame structures are present at 19,578 ft; all of these are in the thinner bed intervals. The core is from Well 470 N. 1.

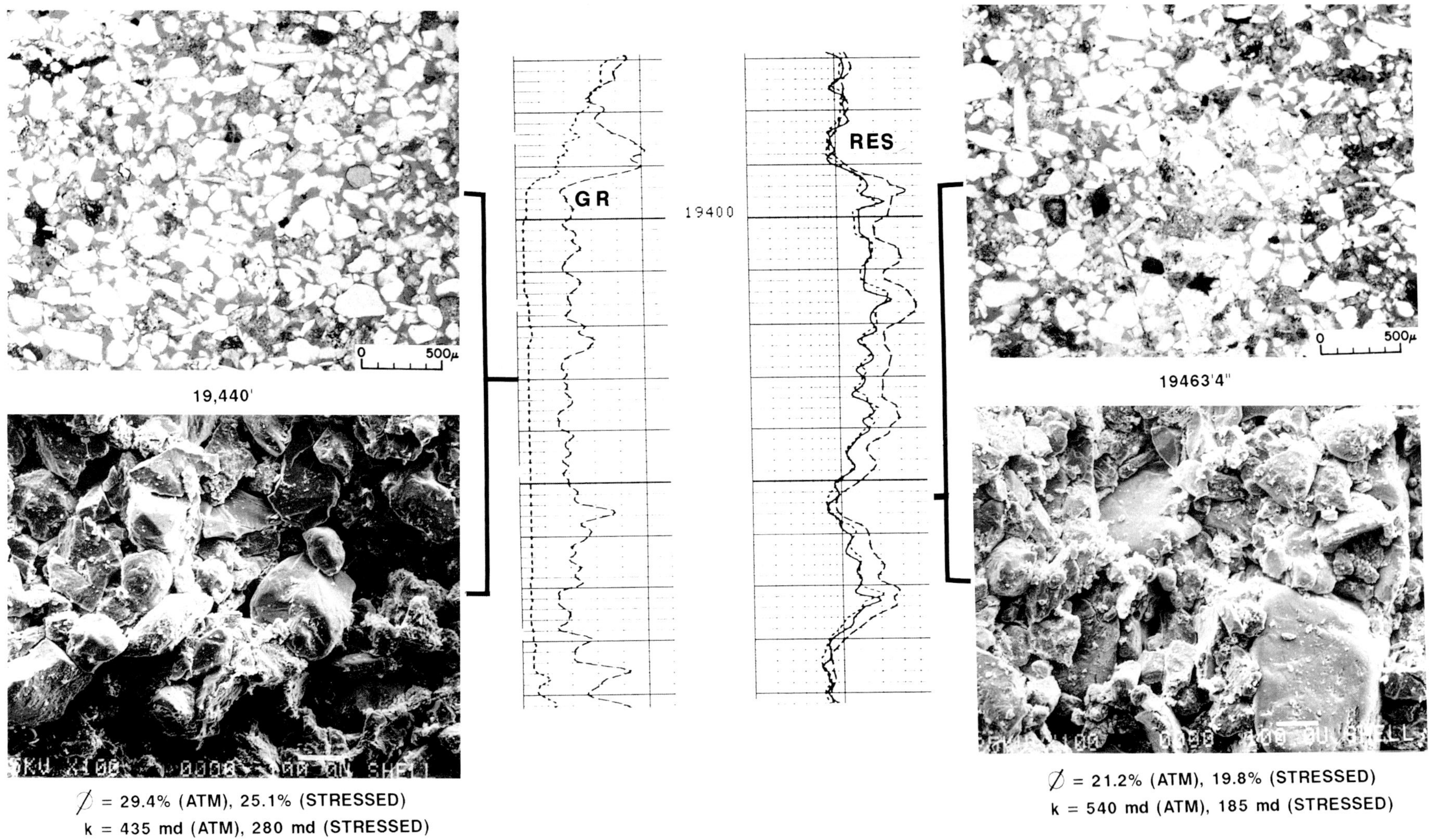

FIG. 11.—A trend of a slightly coarsening upward sequence is obvious from logs and thin sections from the S1 interval in Well 470 No. 1. Note the slight drop in resistivity and "dirtier" gamma ray at 19,450 ft (5928 m). The depths shown on the thin sections are core depths; the core was shifted 10 ft (3 m) to match the log. Porosity and permeability values measured at reservoir stresses are most accurate for these slightly consolidated sands.

areal extent of the "S2" aquifer (Fig. 5). The "S2" sand is expected to receive good pressure support from its aquifer, but the "S1" may need water injection to maintain reservoir pressure. This interpretation will be tested by the production performance of the seven "S" reservoir wells.

RESERVOIR QUALITY AND PRODUCTION

The reservoir petrophysical properties of these highly continuous sheet sands are moderate. The reductions in porosity and permeability are mostly a result of burial and mechanical compaction, but depositional (facies variability vertically and laterally) controls are also important. The large number of grain contacts and occurrence of long grain contacts in these sands are an indication of the high degree of compaction. There is little cement except for local ferroan calcite nodules and siderite. There is some minor grain-coating clay, which is mostly composed of stable, mixed-layer clays with a low smectite to illite ratio. Depositional controls are shown in Figure 11. The slightly coarsening-upward sequences have slightly poorer quality, clay-rich and finer sand at the base grading up to massive cleaner sands at the top. This is illustrated by the resistivity and gamma ray changes shown in Figure 11. The beds are interpreted to be highly continuous based on seismic, well log, and core correlations. Figure 12 illustrates how beds in the two cored wells may be correlated approximately one mile (1.6 km) apart; only slight thinning is evidenced downdip. Figures 13 and 14 illustrate the change in reservoir quality from the more thinly bedded intervals to the massive intervals. The thinner-bedded, depositionally low energy intervals typically have poorer reservoir properties. They are slightly finer-grained, clay-rich, and more poorly sorted with resulting lower porosity, and even more importantly, with lower permeabilities.

The "S" sand contains an estimated 150 MMBO. Six wells were drilled at a relatively wide spacing along the updip margin of the sand. A seventh well (No. A-9) was drilled in a downdip location. Production from the "S" sand began in April, 1994. Currently six wells are producing from the "S" sand. Average individual well rates of 10 MBOPD and 20 MMCFGPD have been established with a peak production of 12.9 MBOPD and 24.7 MMCFGPD from the No. A-7. Total production is currently facilities constrained. Performance will dictate whether pressure support will be necessary.

SUMMARY

The "S" sand reservoir consists of laterally continuous sheet sand deposits formed by a prograding depositional wedge of turbidite-derived sediments. The sediments are part of a ponded basin-fill that onlaps a paleoridge on the northeast side of the basin, which serves as part of the trapping mechanism for the oil. The lateral continuity of the sediments is shown by the sands as well as the mudstones and thin-bedded intervals that can be correlated across the field with well logs and core. The "S" sand can be subdivided into two members, a lower S2 and an upper S1, that are separated by a thick mudstone. The basal part of the S2 is dominated by layered sheet sands, whereas the upper S2 and S1 are dominated by amalgamated sheet sands that occur in coarsening-upward lobe sequences.

Reservoir quality is variable depending upon lithofacies, but within a stratigraphic horizon it remains relatively constant. The massive sands have higher permeabilities and somewhat higher porosities than the thin-bedded intervals that are more clay-rich and finer-grained. The large thickness and continuity of the sands, even with porosities and permeabilities averaging 25% and 150 md, respectively, lead to prolific production, with wells capable of producing an average of 10 MBOPD and 20 MMCFGPD.

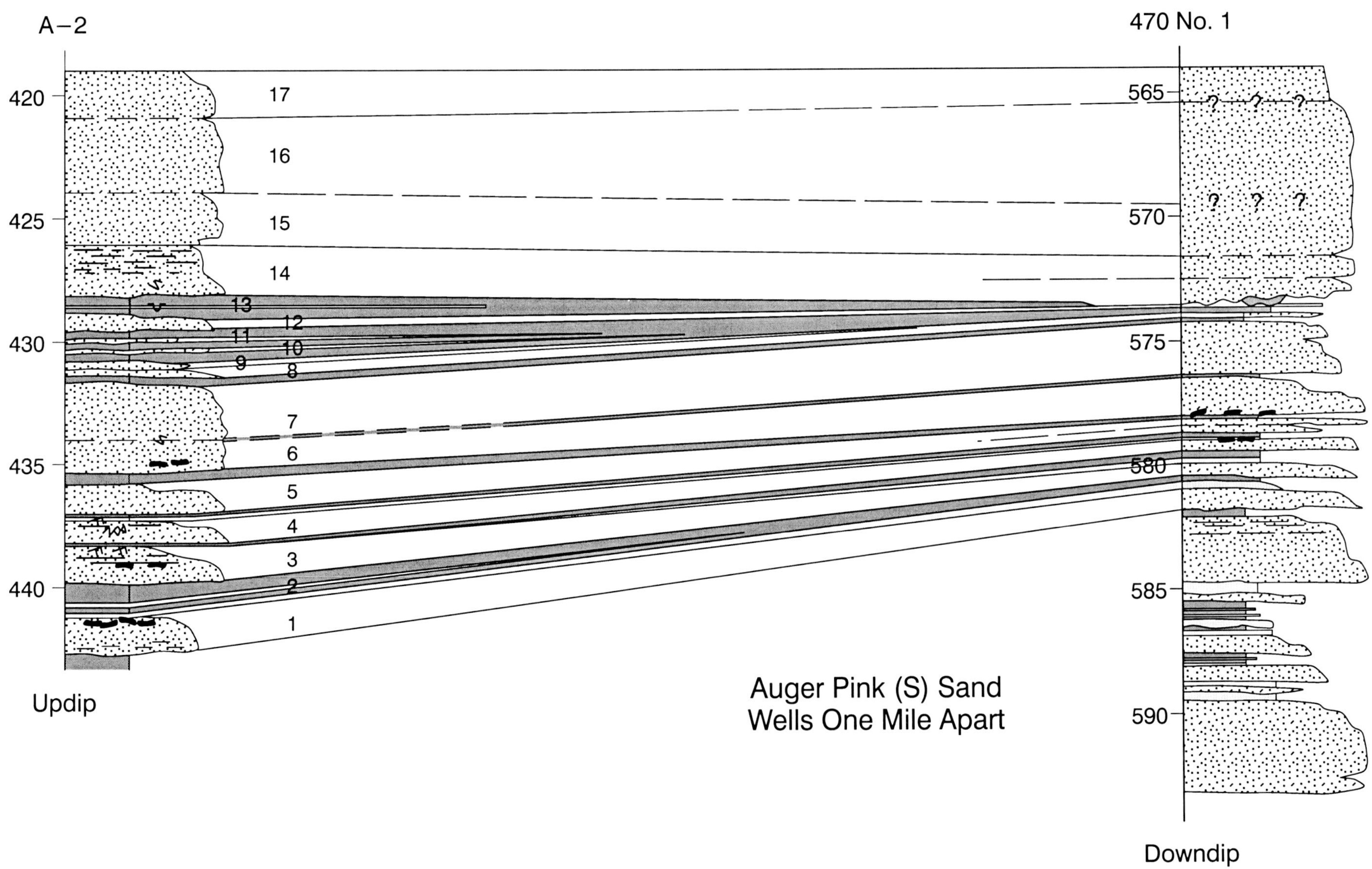

FIG. 12.—Cross section from the updip A-2 core to the downdip 470 No. 1 well. The wells are approximately one mile apart. Although there is some thinning downdip, the sands and intervening mudrocks are highly correlative.

REFERENCES

Bouma, A. H., 1962, Sedimentology of some flysch deposits: Elsevier, Amsterdam, 168 p.

Bilinski, P. W., Crabtree, P. T., Crump, J. G., McGee, D. T., and Watson, R. C., 1992, Auger Field subsurface overview: future site of world-record deepwater tension leg platform: 67th Annual Technical Conference and Exhibition of the Society of Petroleum Engineers, Proceedings, paper number 24701, Washington D.C., p. 387-397.

McCain, D. W. and Bridges, B., 1994, Volatile oils and retrograde gases – What's the difference?: Petroleum Engineer International, January: p. 35-36.

Mutti, E. and Nilsen, T. H., 1981, Significance of intraformational rip-up clasts in deep-sea fan deposits: Abstracts Volume, 2nd European Regional Meeting International Association of Sedimentologists, Bologna, p. 117-119.

Mutti, E. and Normark, W. R., 1987, Comparing examples of modern and ancient turbidite systems: Problems and concepts, in Leggett, J. K. and Zuffa, G. G., eds., Marine Clastic Sedimentology, Chapter 1, Graham and Trotman Press, p. 1-38.

Mutti, E. and Ricci Lucci, F., 1972, Turbidites of the Northern Appennines: introduction to facies analysis (English translation by T. H. Nilsen, 1978): International Geology Review, v. 20, p. 125-166.

Mutti, E. and Sonnino, M., 1981, Compensation cycles: a diagnostic feature of turbidite sandstone lobes: Abstracts Volume, 2nd European Regional Meeting International Association of Sedimentologists, Bologna, p. 120-123.

Normark, W. R., 1978, Fan valleys, channels, and depositional lobes on modern submarine fans: characters for recognition of sandy turbidite environments: American Association of Petroleum Geologists Bulletin, v. 62, p. 912-931.

Shew, R. D., Gary, P. S., Pfeiffer, D. S., and Bilinski, P. W., 1993, Evidence and importance of a carbonate-rich deposit formed in association with a paleo-hydrocarbon seep: American Association of Petroleum Geologists, Annual Convention Program with Abstracts, New Orleans, p. 118.

Steffens, G. S., 1993, Gulf of Mexico deepwater seismic stratigraphy: American Association of Petroleum Geologists, Annual Convention Program, New Orleans, p. 186.

Walker, R. G., 1978, Deep-water sandstone facies and ancient submarine fans: models for exploration for stratigraphic traps: American Association of Petroleum Geologists Bulletin, v. 62, p. 932-966.

RAM/POWELL FIELD: VIOSCA KNOLL BLOCK 912, DEEPWATER GULF OF MEXICO

GEORGE R. CLEMENCEAU
Amoco Production Company, Offshore Business Unit, New Orleans, Louisiana

ABSTRACT

Ram/Powell field is an example of a passive margin deep-sea fan stratigraphic trap. Deposition of the Ram fan was controlled by sea level lowstand, intraslope basin geometry, and a point-source deltaic sediment supply. Multiple elongate lens-shaped fanlobes have been interpreted using 3D seismic, conventional-core, and well log data. Seismic reflection clinoform configurations indicate lithofacies distribution within the fanlobes. The reservoirs are trapped by stratigraphic pinch out of channel and levee-overbank sands. The best quality reservoirs are associated with channel-levee deposits, that are dominated by thick, fine-grained, massive turbidite sands.

INTRODUCTION

In the last ten years, substantial hydrocarbon resources have been discovered in the Deepwater Gulf of Mexico, trapped within various types of clastic deep-sea fan systems. The Ram/Powell Field is an example of a stratigraphic trap within a slope-fan system. At Ram, reservoirs occur in both channel and levee-overbank sands, but the major productive interval is the channel facies. Representative slope-fan facies are described below using conventional core interpretations integrated with well log and 3D seismic data.

Ram/Powell Field is in the Offshore Northern Gulf of Mexico (GOM), 50 miles east of the Mississippi River Mouth, in 3200' of water (Figure 1). Amoco, Exxon, and Shell are the working interest owners of this 1985 Shell discovery. The initial drilling found gas pay associated with a seismic amplitude anomaly. Confirmation drilling established four major hydrocarbon-zones between 12300' and 13100' (subsea depth) (Figure 2). Amoco estimates that the field contains ~300 MMBOE. Development drilling will begin in 1995.

GEOLOGICAL SETTING

Ram/Powell Field is stratigraphically trapped within an upper middle Miocene deep-sea fan complex that was deposited in a passive margin intraslope basin setting. In general, deep-sea fan architecture is controlled by sea level change (eustatic or tectonic), regional setting (basin geometry and bathymetry), and sediment supply (delivery system, climate, and sediment type)(Stow et al., 1983; Nelson and Nilsen, 1984). Deposition of the Ram fan was controlled by sea level lowstand, intraslope basin geometry, and a point-source deltaic sediment supply. Fan deposition is approximately synchronous with a lowstand event at ~11 Ma on the Haq (1987) sea level curve, according to the age of the associated marker fossils (Figure 3). The fan was deposited in a water depth of 500 to 6000 feet, according to foraminifera paleobathymetry. Turbidity currents, as indicated by core interpretation, transported sand downslope through a single feeder channel into an intraslope basin 50 miles southeast of the middle Miocene shelf margin (Figure 4).

Salt domes and ridges frequently control the shape of intraslope basins in the Gulf of Mexico, thereby strongly influencing fan geometry. However, according to seismic interpretation, salt movement in this area occurred after deposition of the Ram fan. The lack of physiographic restrictions during deposition allowed the fan to aggrade and shift laterally with time rather than stacking vertically as in more restricted basin settings (Figure 5). Because of its unrestricted setting, the Ram fan architecture is unusual in comparison to most other Gulf of Mexico Deepwater discoveries.

CARDIFF
UWCC LIBRARY

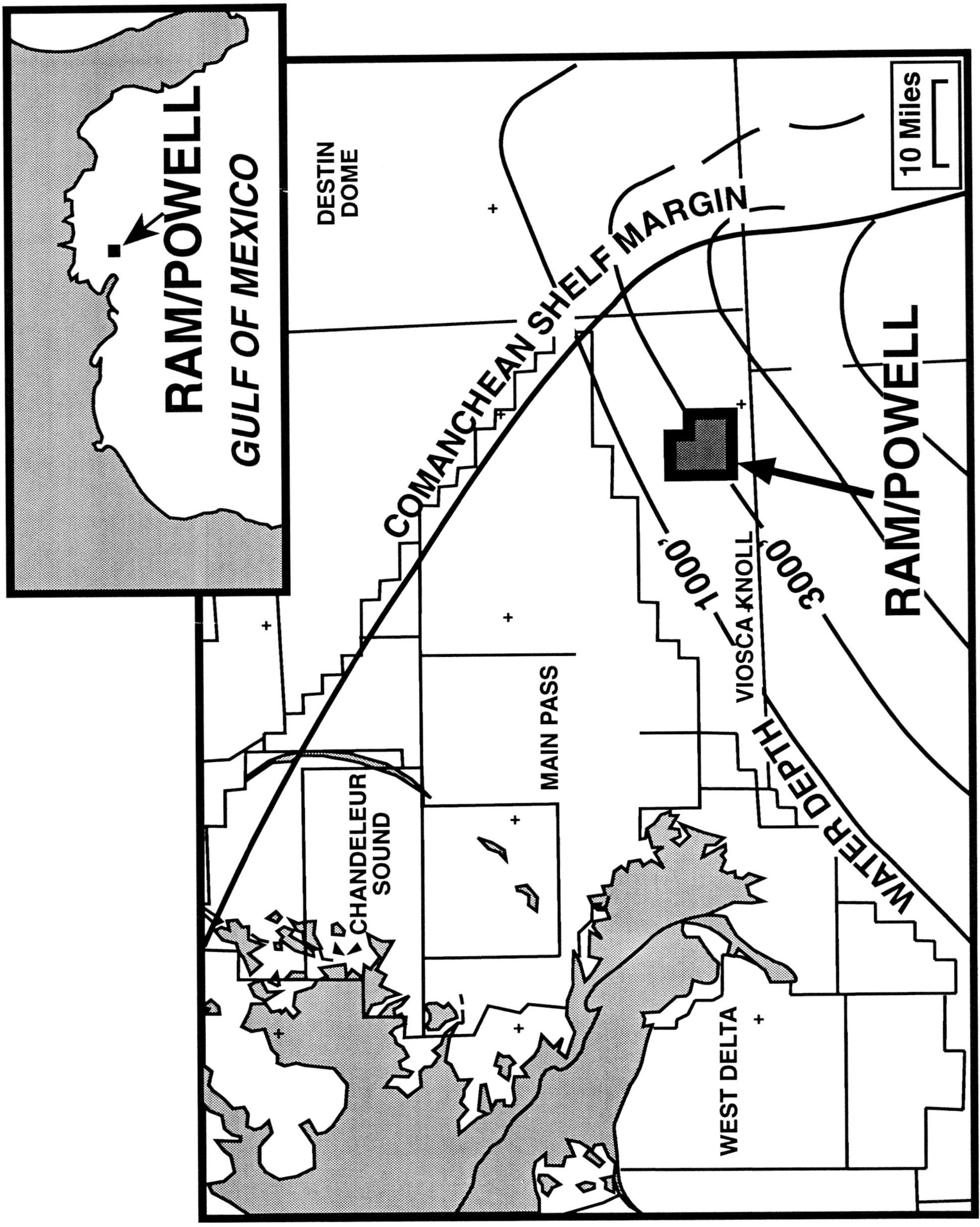

FIG. 1.—Ram/Powell Field Deepwater Gulf of Mexico.

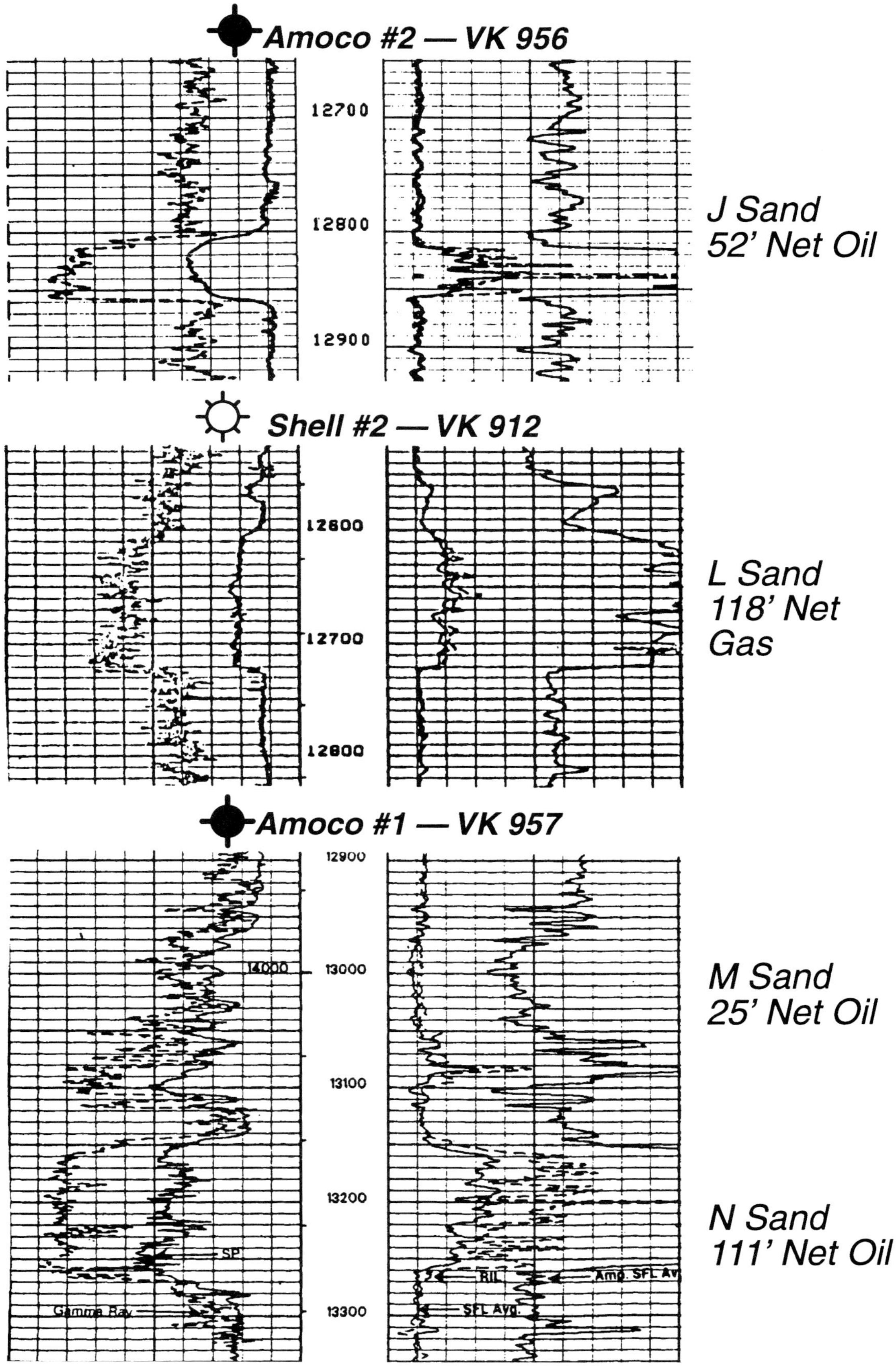

FIG. 2.—Ram/Powell Field major pay sands.

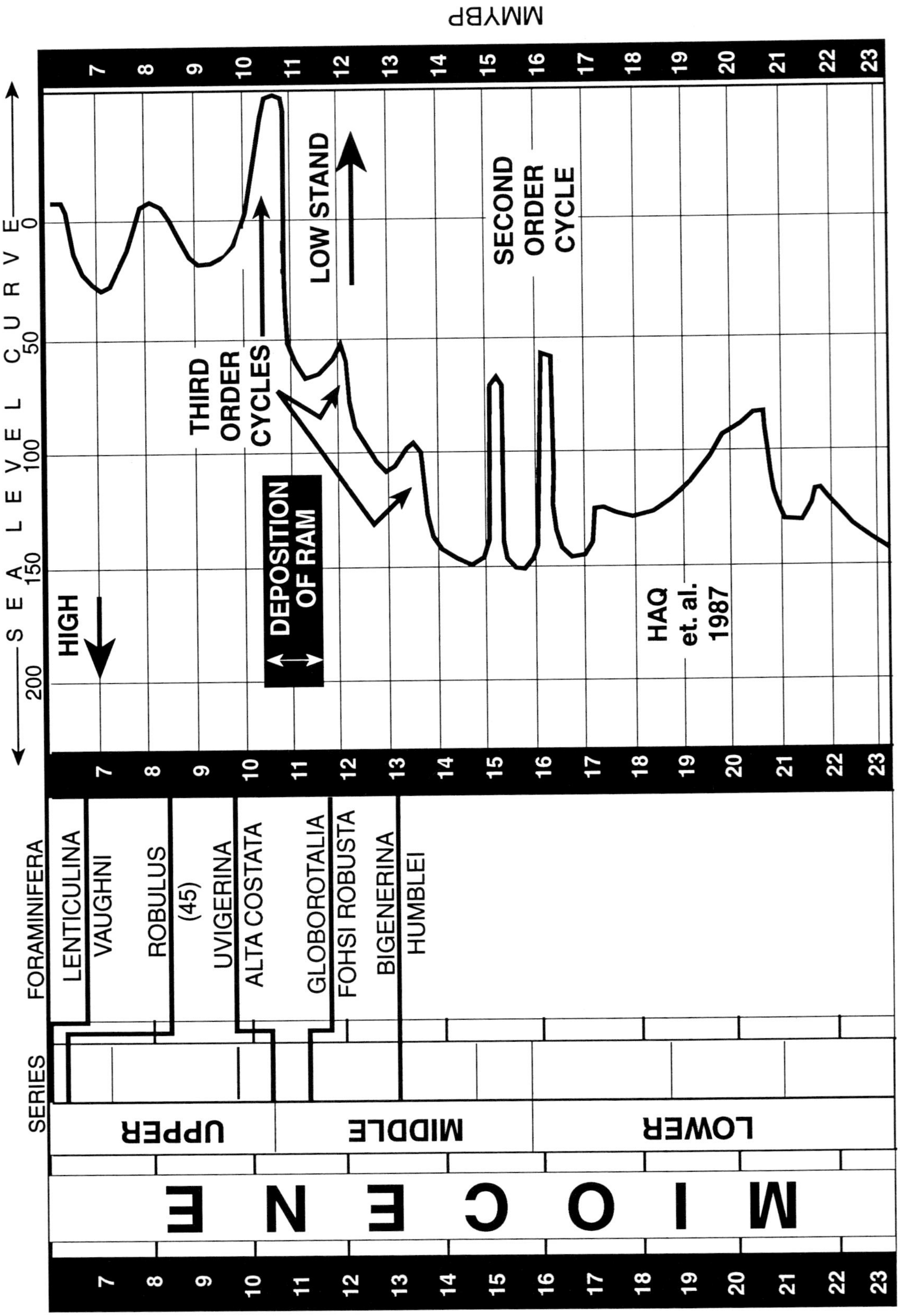

FIG. 3.—Relationship of Ram deposition to sea level.

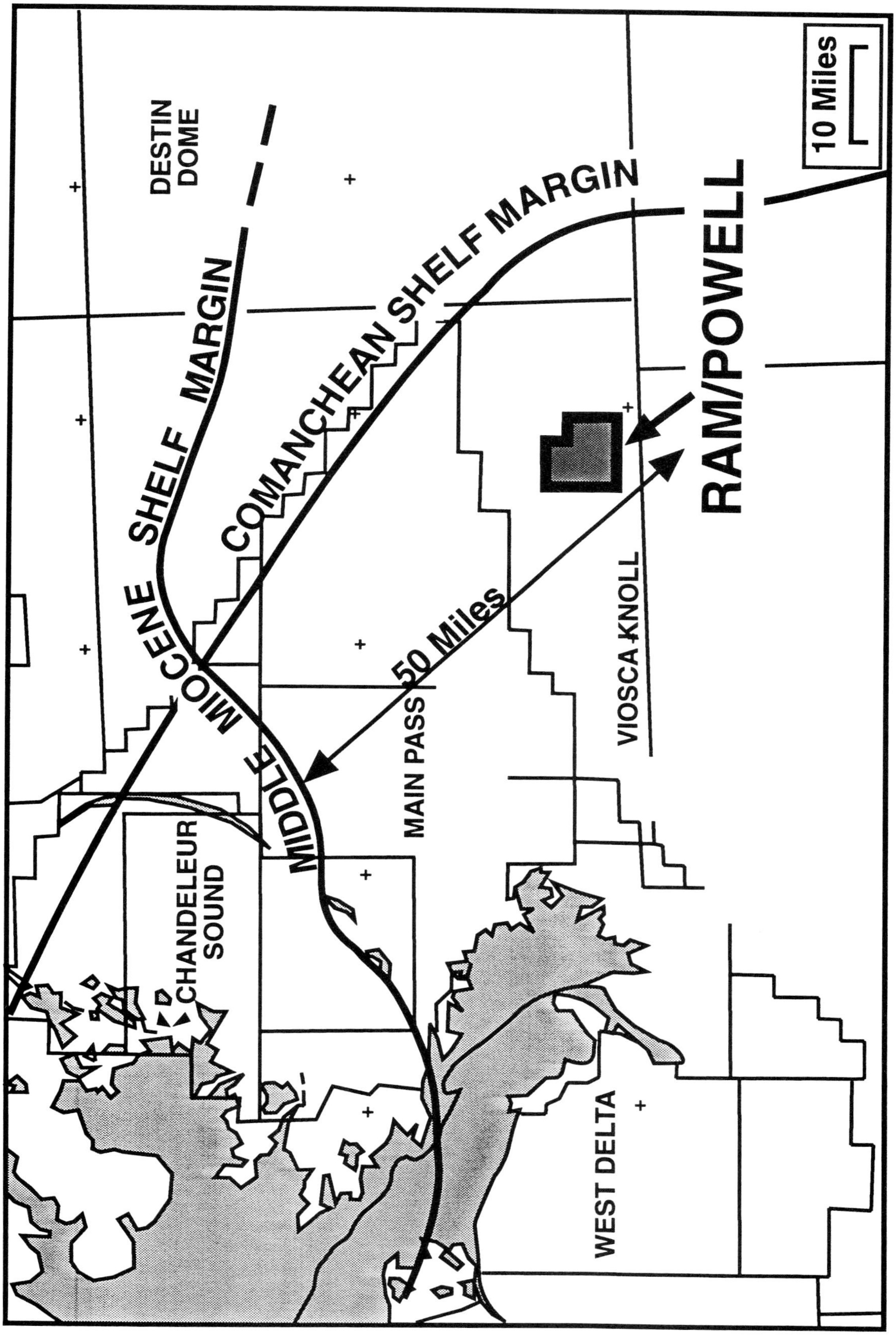

FIG. 4.—Relationship to the middle Miocene shelf margin.

FAN ARCHITECTURE

The Ram slope fan system is composed of eight (or more) elongate fanlobes, as defined by 3D seismic amplitude and isochron map interpretations integrated with well log data (Figure 6). Lobes not penetrated by wells have seismic amplitude characteristics similar to the logged pay sands and are considered sand prone. All the fanlobes are in a northwest-southeast orientation, which suggests a point source feeder channel to the north, outside of the 3D seismic data coverage. The oldest lobes (X and Y, Figures 6 and 7) are located in the northeast section of the field. Upsection, the fan shifts laterally to the southwest (lobes N to J), then migrates back to the northeast (lobes G2,G1) onlapping previously deposited lobes (Figure 7). The lateral shifting of deposition results in a fan-shaped deposit covering an area of approximately 50 square miles. Within the field, fanlobes vary from one to three miles wide and up to nine miles long. Though most of the fanlobes are 50-80 feet thick, they reach a maximum gross thickness of 200 feet, as indicated by seismic data. The average logged gross thickness is 70 feet. The thickest sands are associated with the axis of each lobe.

Lithofacies distribution within fanlobes is interpreted from internal-external seismic reflection clinoform configurations as described by Sangree and Widmier(1977). In strike-section view, the lobes are broadly mounded or convex-up lens-shaped in form (Figures 8 and 9). In each fanlobe a stratigraphic trap is formed by sand pinch out, up dip and laterally, into hemipelagic shale. The shingled clinoform configuration within the upper and middle section of the J fanlobe (Figure 10, line A) changes to a continuous/parallel configuration in the lower section (Figure 10, line B). This is interpreted as a meandering channel deposit amalgamating to sheet sands in a downcurrent direction. The L fanlobe has a lens-shaped continuous configuration interpreted as lower energy deposition consistent with a levee-overbank environment (Figure 11). Figure 11 illustrates the lateral migration and aggradation of the M, L, and J fanlobes. The N fanlobe (Figure 12), which has an irregular base suggesting erosion, and an internal fill clinoform configuration, is interpreted as an erosional channel fill deposit.

In summary, the distinctive features of the Ram/Powell fan include a lobate to elongate shape, meandering and laterally migrating lenticular channel-levee deposits, and a moderate to high sand/shale ratio. These features are characteristic of a single-source mud/sand-rich fan system as described by Reading and Richards (1994).

CORE INTERPRETATION

The major Ram/Powell reservoirs are trapped within massive sands associated with inner and mid-fan depositional environments, according to an interpretation of the conventional cores. The cores, recovered from all major reservoirs, are representative of deep-sea fan channel and levee-overbank facies sands. These sands are thickest in the channel axes and thin laterally to interbedded levee-overbank sand, silt, and shale. The channel facies cores are dominated by facies B turbidites as described by Mutti and Ricci Lucchi (1978), whereas facies C, D, and E turbidites dominate cores from the levee-overbank environment.

J Fanlobe Channel-levee Sands

The J sand structure and isopach map (Figure 13) shows the variation in thickness of this reservoir sand and the location of the two wells selected for discussion. The core recovered from the middle of a 53' oil-pay in the VK 956 Amoco #2 well is an example of the J channel facies (Figure 14). This channel interval is composed of massive, fine-grained, moderately sorted sand. Little or no grading and few sedimentary structures are evident. Thin silt and shale laminations are noted in places. Figure 15 shows the characteristic features of this core. These massive tabular bedded sands with rare shale laminations are interpreted as Mutti facies B turbidites (Mutti and Ricci Lucchi, 1978).

The VK 912 Shell #2 well drilled the J sand outside the channel axis (Figure 13) and encountered a 33' levee-overbank facies sand that contains 17' of net gas pay (Figure, 16). The 30' core recovered from this sand is composed of a stacked sequence of Mutti facies C2 turbidites. The sand is massive at the bed base with an erosional contact. The interval fines upward to parallel and ripple laminated sand, then changes to thin silt and shale laminations (Figure 17). Some finely laminated sand is burrowed. The upper 10' of section (Figure 18) contains Mutti facies D and E turbidites and syndepositional slumping characteristic of levee-overbank deposition (Nelson and Nilsen, 1984).

Integration of core and seismic interpretations suggests that the J fanlobe represents mid-fan channel-levee facies deposition. The massive Mutti facies B sands are associated with the thick sinuous fanlobe axis, as defined by seismic mapping (Figure 13). These sands grade laterally to thinning upward Mutti C2, D, and E turbidites interpreted as levee-overbank deposits. The levee-overbank sands pinch out laterally into shale, as shown by seismic sections (Figure 10).

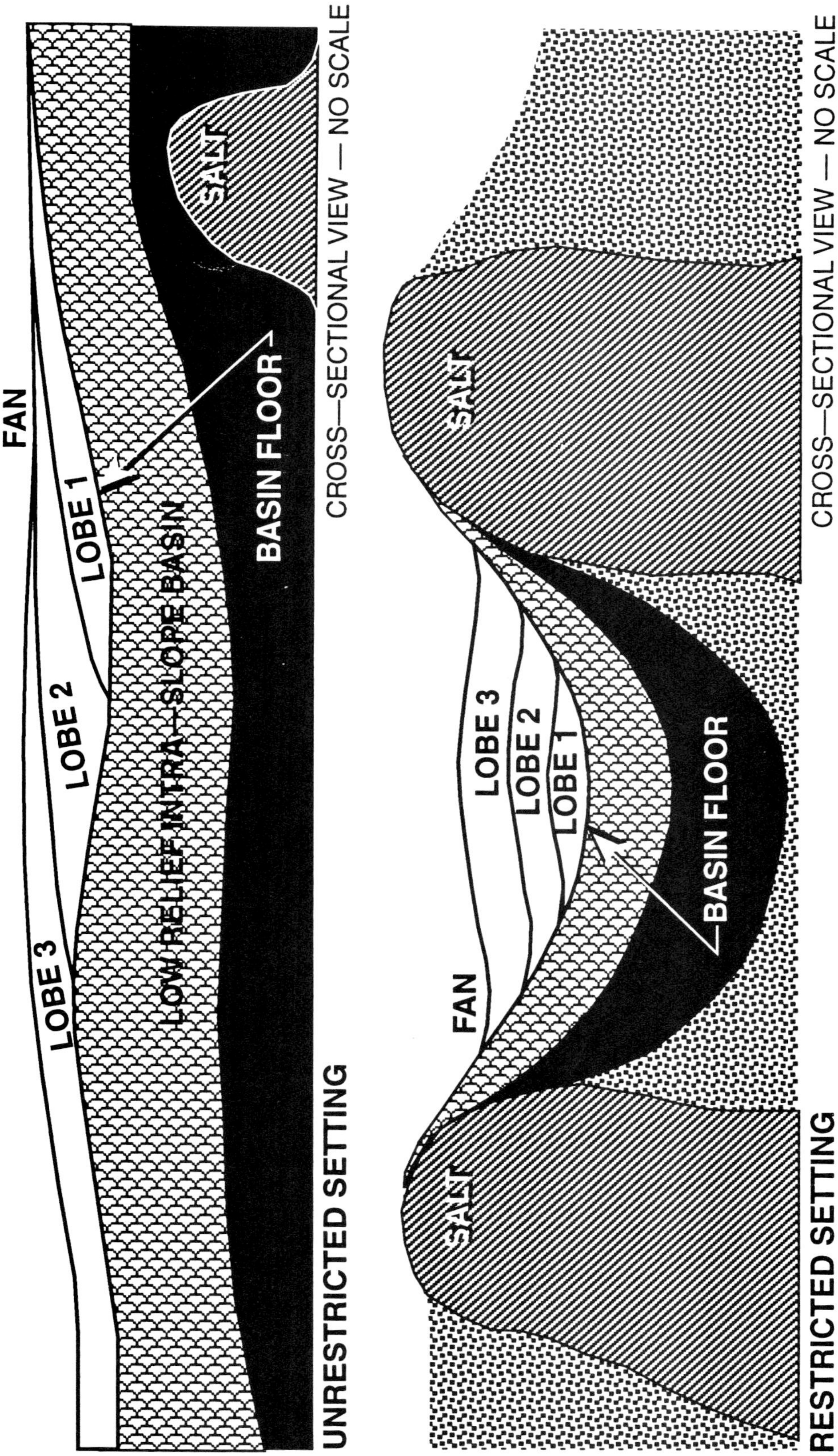

FIG. 5.—Conceptual comparison of fan architecture.

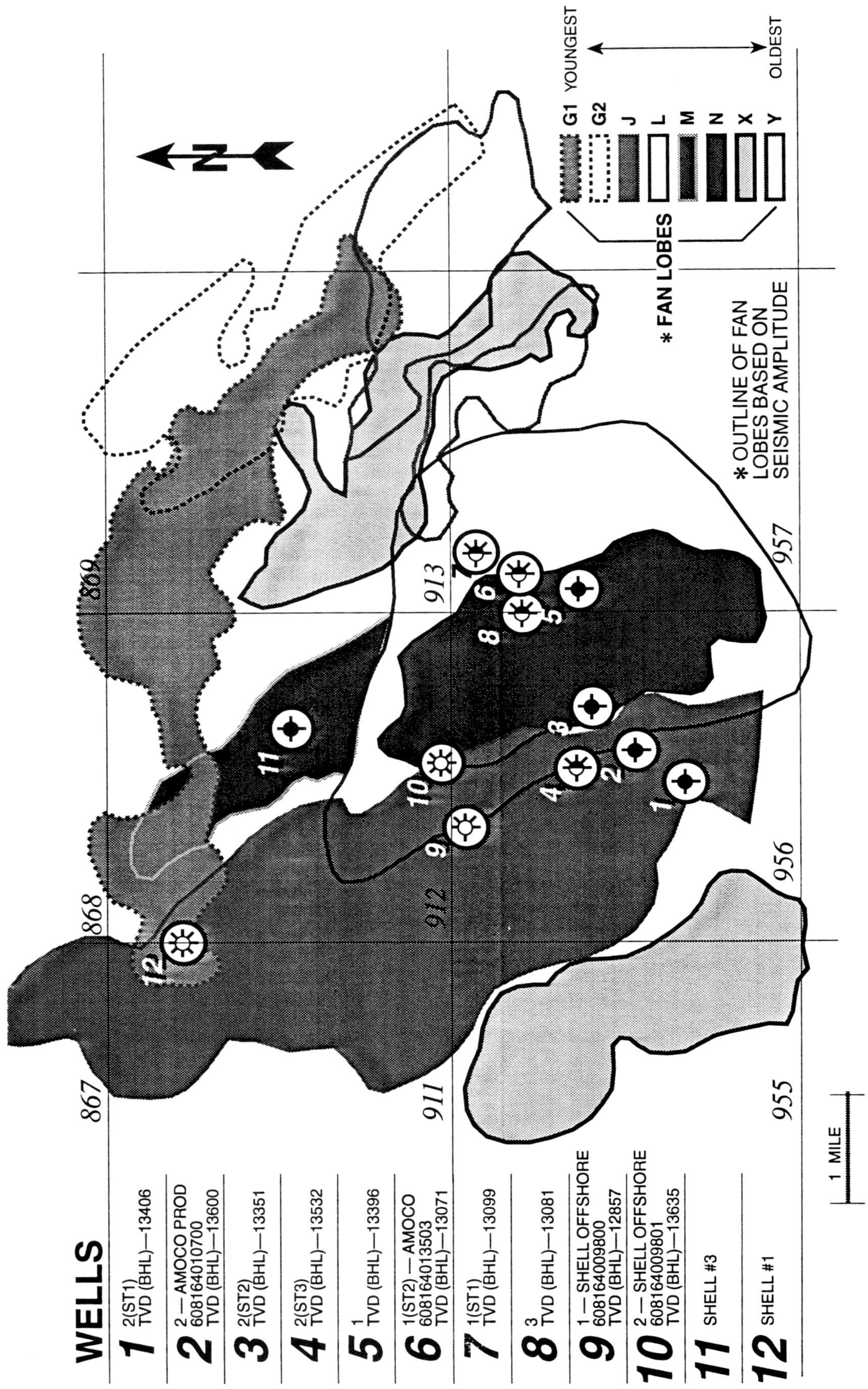

FIG. 6.—Fanlobe Composite Map.

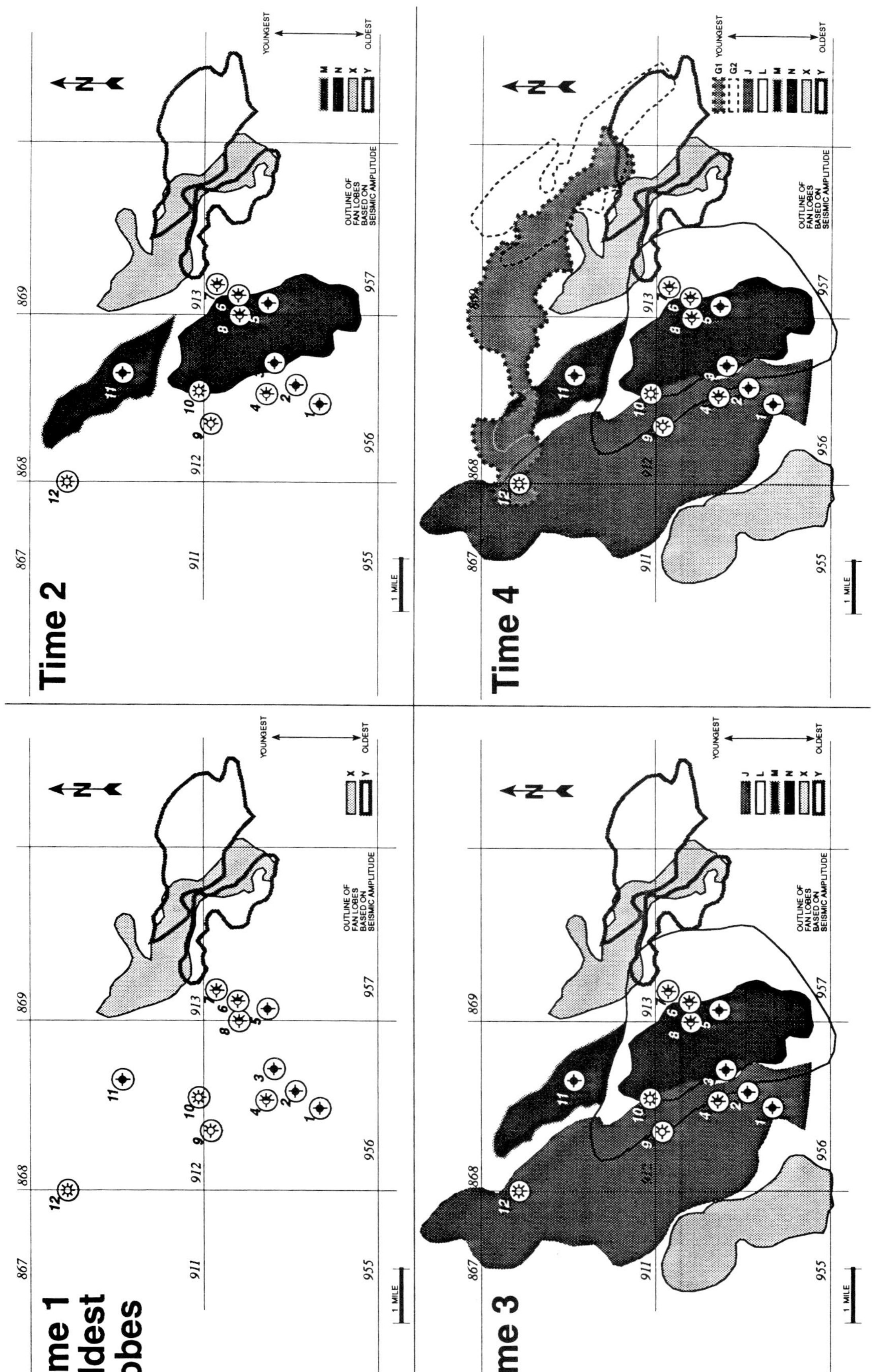

FIG. 7.—Lateral aggradation of Fanlobes.

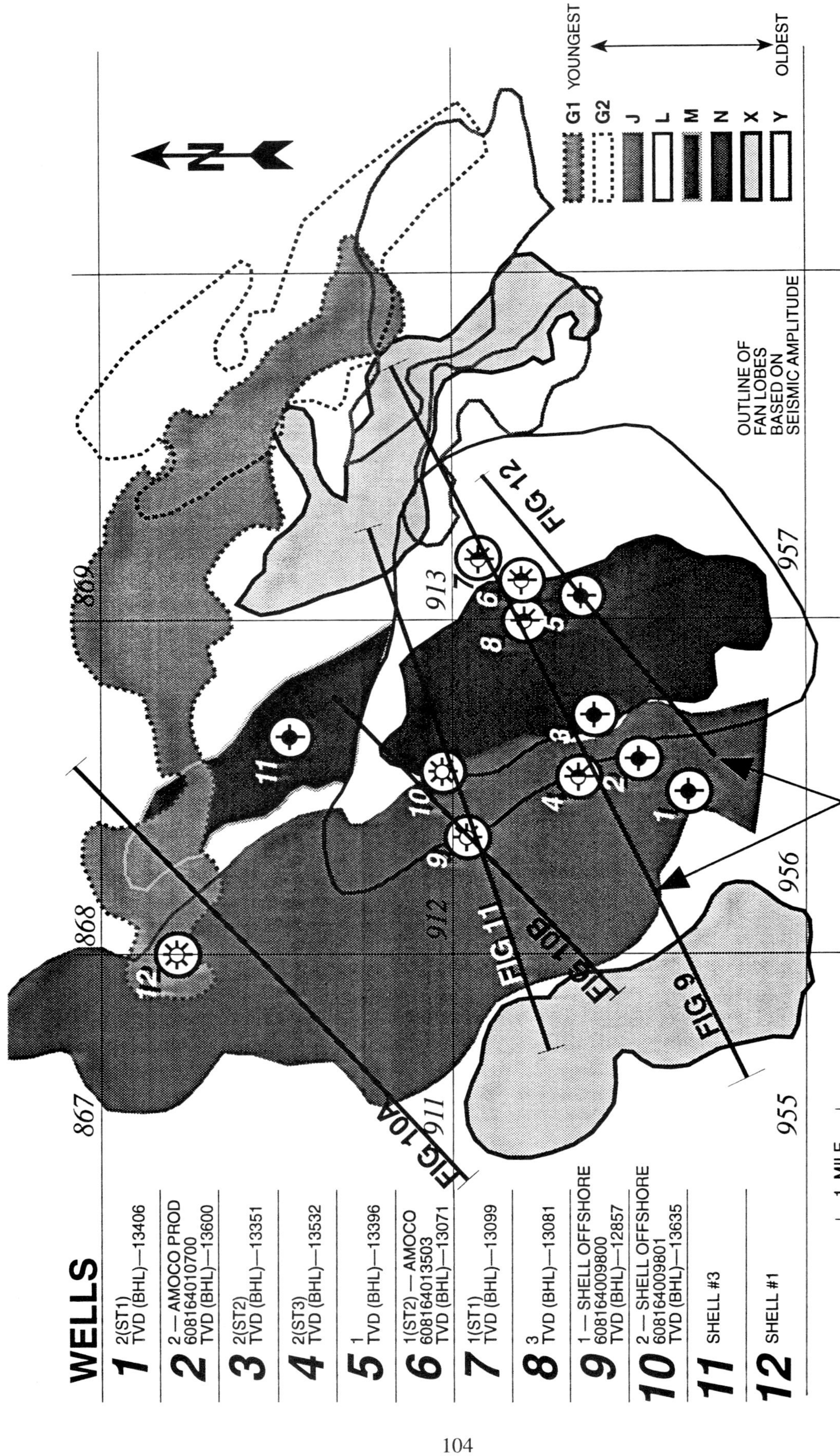

FIG. 8.—Fanlobe composite map with seismic line locations.

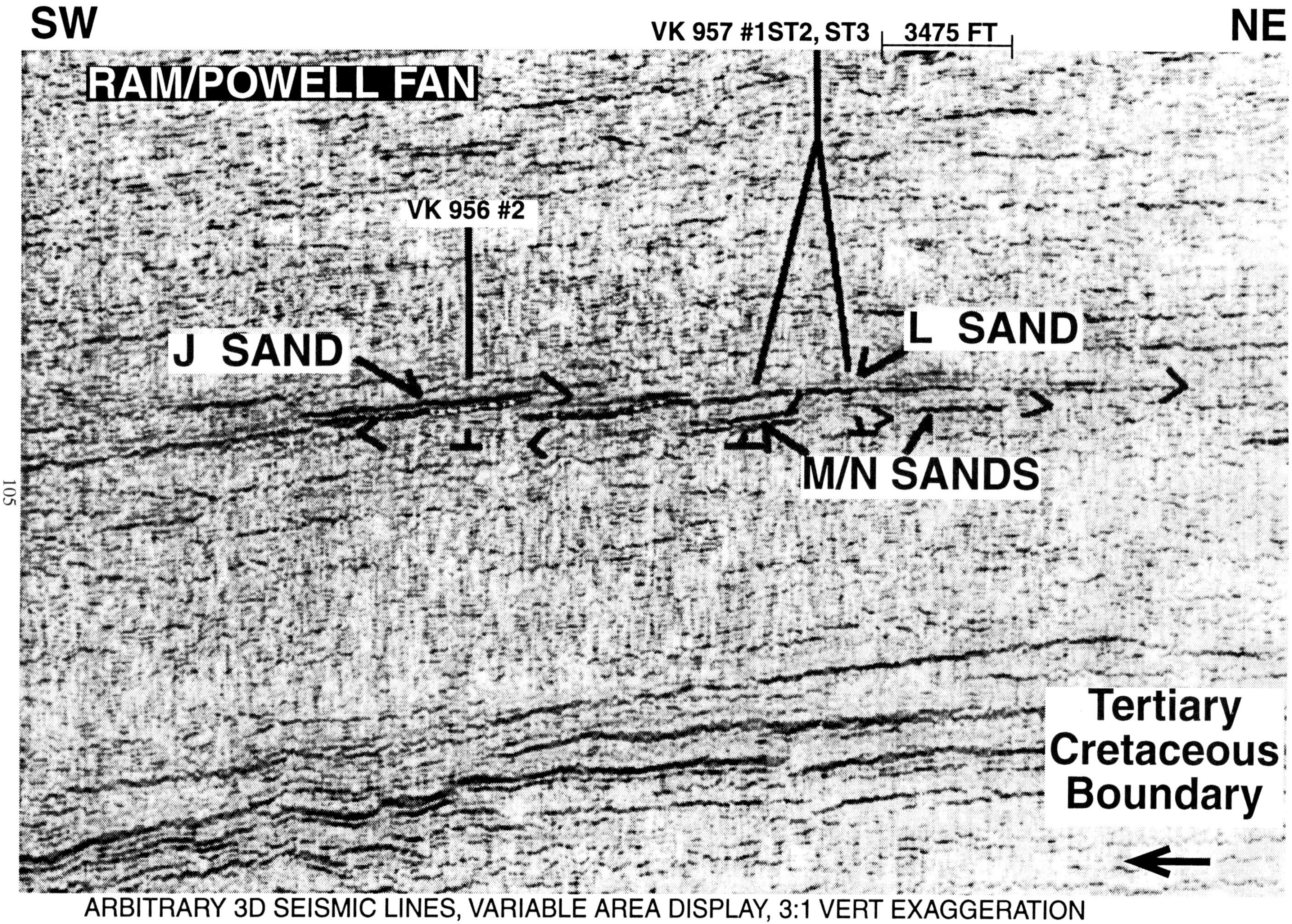

FIG. 9.—SW-NE 3D seismic line showing the fan architecture in strike section view.

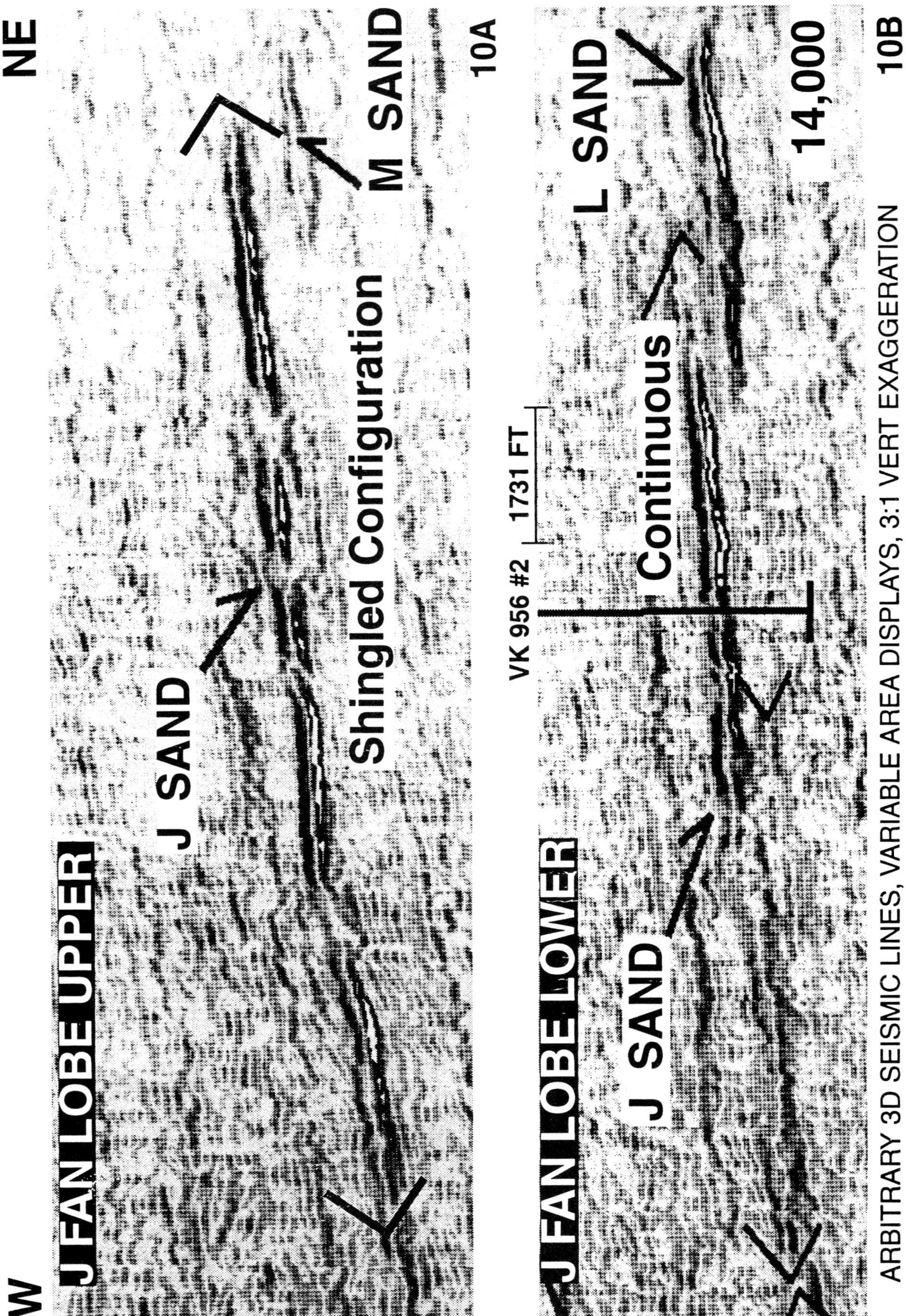

FIG. 10.— SW-NE 3D seismic line comparing seismic clinoform configuration in the upper J fanlobe to the lower J fanlobe.

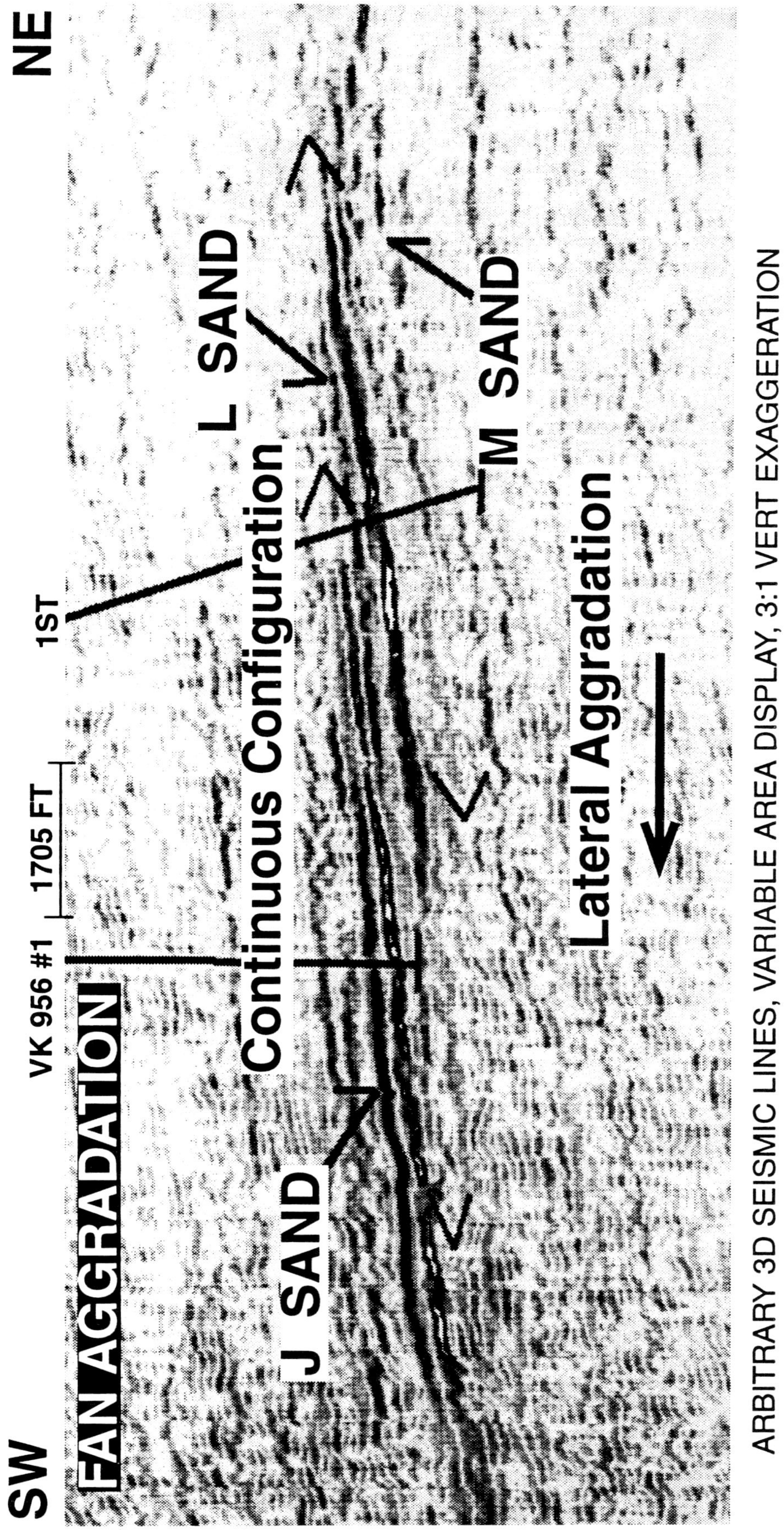

FIG. 11.— SW-NE 3D seismic line showing the aggradation and lateral shifting of fan deposition with time.

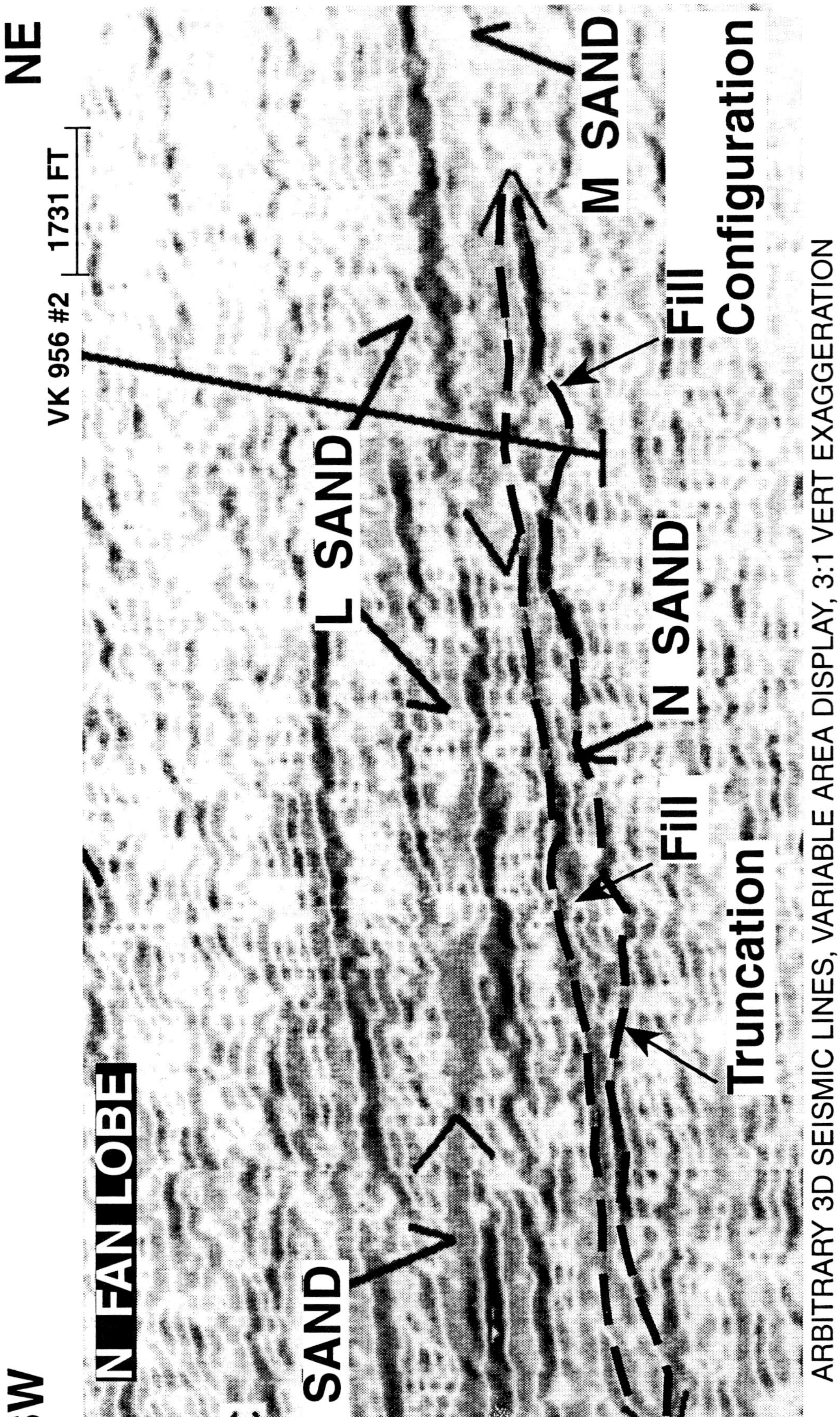

FIG. 12.—SW-NE 3D seismic line showing the erosional base and fill clinoform configuration of the N sand fanlobe.

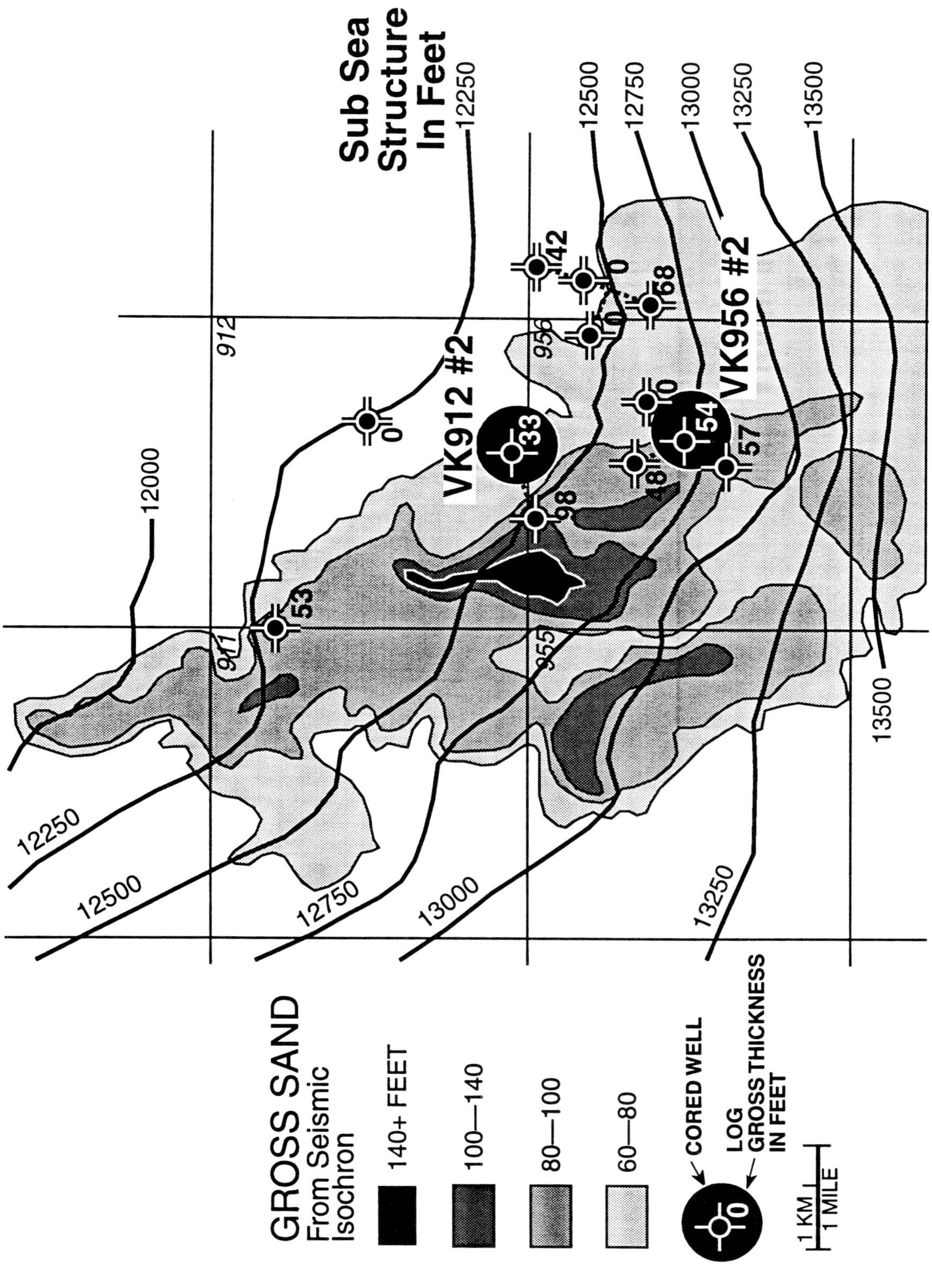

FIG. 13.—J fanlobe gross sand isopach with structural contours.

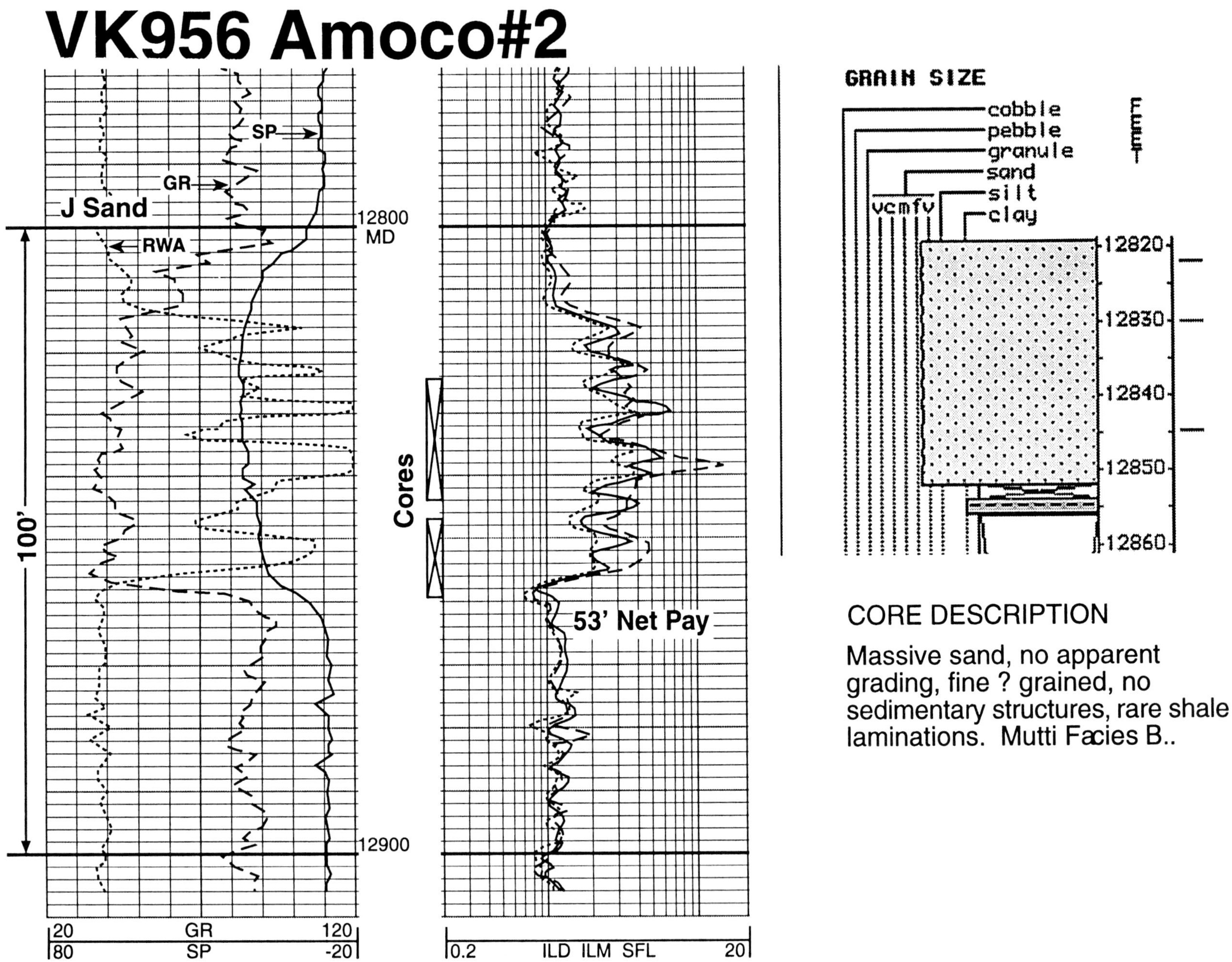

FIG. 14.—J sand mid-fan channel facies. The cored interval is a fine grained massive sand with no apparent grading or sedimentary structures. This interval is interpreted as a Mutti facies B turbidite.

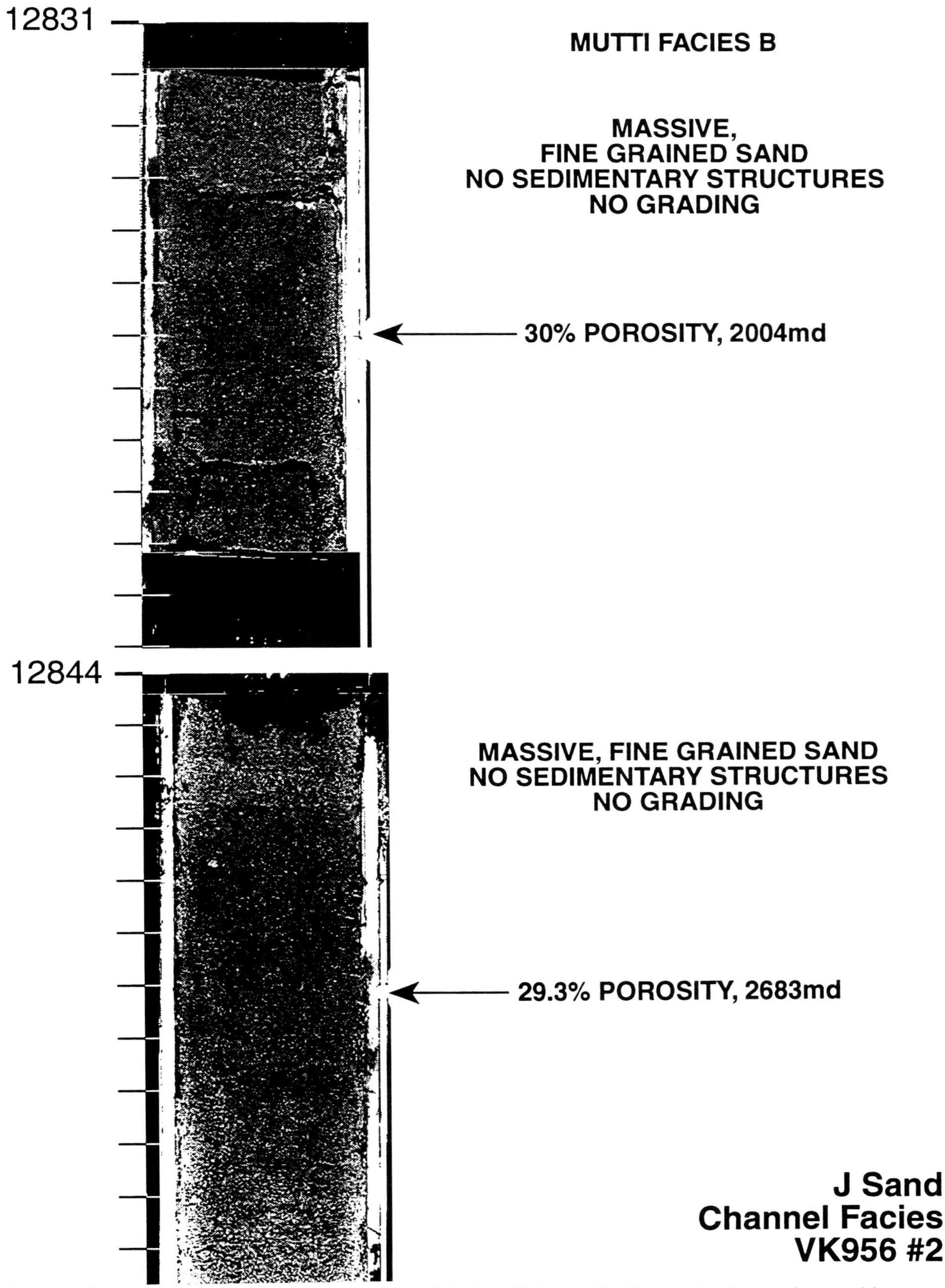

FIG. 15.—Core photograph of the J channel sand facies. This sand is fine grained, massive, and has no apparent sedimentary structures. It has a porosity of 29-30% and a permeability of 2004-2683mD.

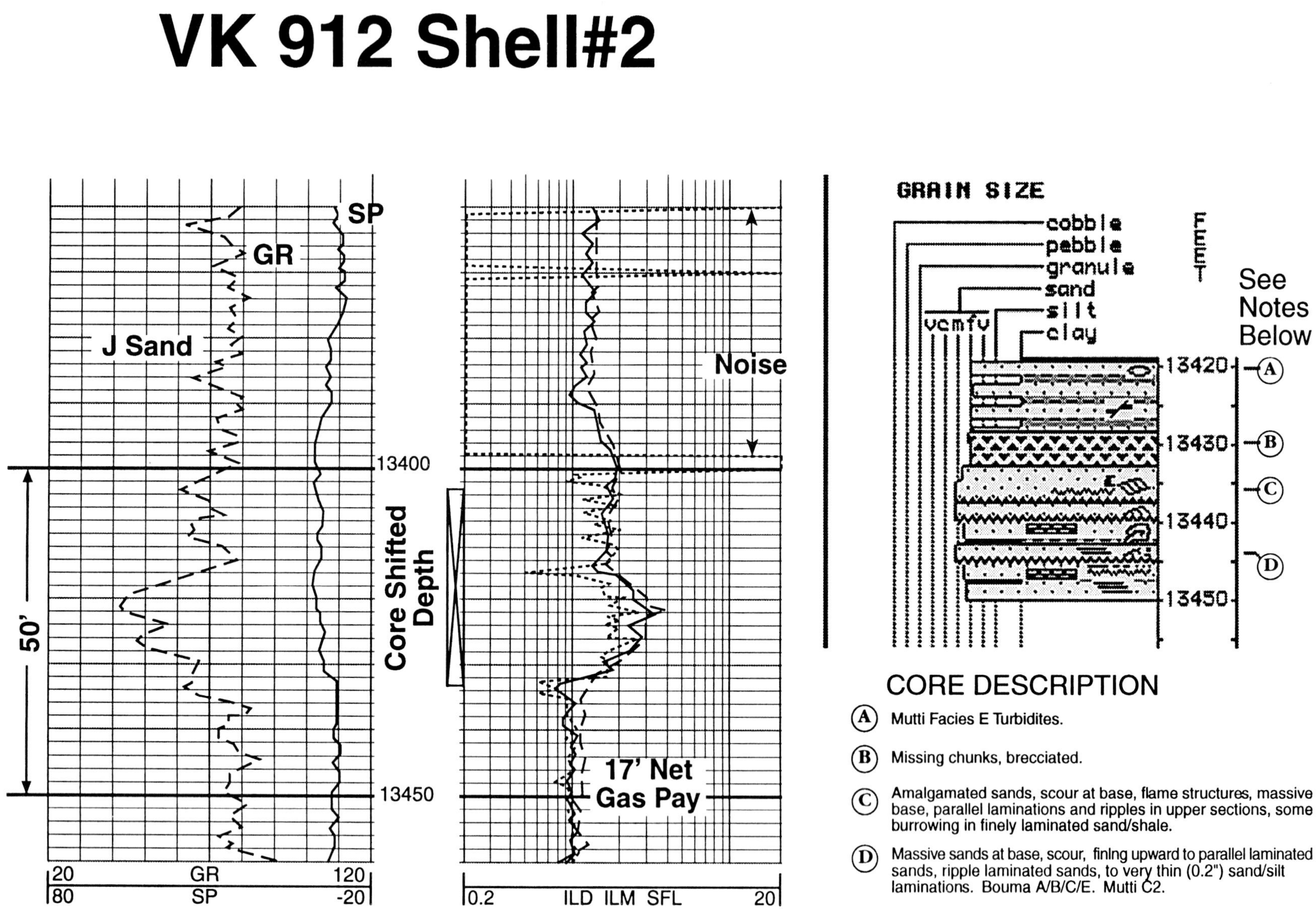

FIG. 16.—J sand levee-overbank facies. The cored interval is interpreted as Mutti facies C2 and facies E turbidites.

N Fanlobe Channel Sands

The N sand is interpreted as an inner-fan erosional channel-fill because of its rapid lateral pinch out, seismic fill configuration (Figure 12), and Mutti facies B core characteristics. This fanlobe is narrower in map view and thicker than the J fanlobe discussed above (Figure 19). The only core obtained from this sand was the thirty-two feet of core recovered from the VK 956 #3 well (Figure 20). Like the J channel sand above, the N sand is a massive medium-grained Mutti facies B turbidite with no apparent sedimentary structures (Figure 21). Thin parallel shale laminations are noted in places. Other wells, such as the VK 957#1 well (Figure 22), although not cored, logged a thick blocky-character sand that could also be interpreted as Mutti facies B.

L Fanlobe Levee-overbank Sands

The L fanlobe is interpreted as a levee-overbank deposit. Figure 23 shows the location of two wells selected for discussion, the VK 957 #1st2 and the VK 912 #2. Cores from these two wells represent the variation of the L sand levee-overbank facies characteristics. (Note that the map is an outline of the seismically defined reservoir area, not an isopach.)

The #1st2 well logged 20' of net gas pay in a 66' gross thick sand (Figure 24). Cores from the pay sand contain sequences of very-fine-grained thinning and fining upward turbidites. Individual beds are characterized by a sharp-based, parallel laminated interval overlain by ripple laminated sands and capped by thin interbedded sand, silt and shale (Figure 25). These L sand cores are interpreted as Mutti facies D1 and D2 turbidites. Also recognized are lenticular, or flaser-bedded, ripple-laminated sands interpreted as Mutti facies E turbidites (Figure 26). Climbing ripples, burrows, and soft sediment deformation features that are characteristic of the levee-overbank environment are present throughout (Walker, 1992; Nelson and Nilsen, 1984).

The VK 912 #2 well logged a 72' net pay in a relatively thick, 120' gross, L sand interval (Figure 27). The cores recovered in this sand are composed of very-fine-grained, thinly laminated (<1') Mutti facies D and E turbidite sands interbedded with shale. Climbing-ripples, convolute, lenticular, and wavy parallel beds are present throughout the cores. Rip-up clasts and scours are also noted. Soft sediment slumps and Mutti facies E lenticular ripple-laminated beds are common (Figures 28, 29, 30).

RESERVOIR QUALITY

Ram/Powell reservoir sand quality is excellent. Conventional-core porosity for the field ranges from 19 to 32%. Core permeability (K air) ranges from ~10 to ~3000 mD. The channel sands, which are the thickest and have the highest net/gross sand ratio (net sand >50%), have excellent porosity and permeability. The J channel sand averages 30% porosity and 1600mD permeability (K air). Permeability ranges from 700 to 2683 mD. The L sand levee-overbank turbidites average less than 50% net sand. Individual laminations in these thin bedded sands typically have about the same porosity as the channel sands, but permeability is lower due primarily to finer grain size. L sand permeability ranges from ~10 to ~1300 mD.

The slumped, convolute, and lenticular bedding observed in cores of the Ram levee-overbank sands suggests that reservoir continuity may be poor. However, a four day production test of a levee-overbank sand containing 17' of net pay, in the VK 957#1 well, flowed at 22 MMCFPD and 2670 BCPD (46/64" choke, 2800 FTP). The reservoir permeability indicated by this test is 375 mD, which is consistent with the core data discussed above. The flow test suggests that reservoir continuity is better than indicated by cores.

ACKNOWLEDGMENTS

The author wishes to thank the Amoco Production Company and partners Shell and Exxon for permission to publish this paper. The author would like to recognize several individuals whose earlier work contributed to this paper. Patti Miller provided the seismic lines and interpreted the maps used in this study. The cores were originally described by John Laird and Norman Haskell. Gerard Simms and Ken Rockwood studied the core petrophysical properties.

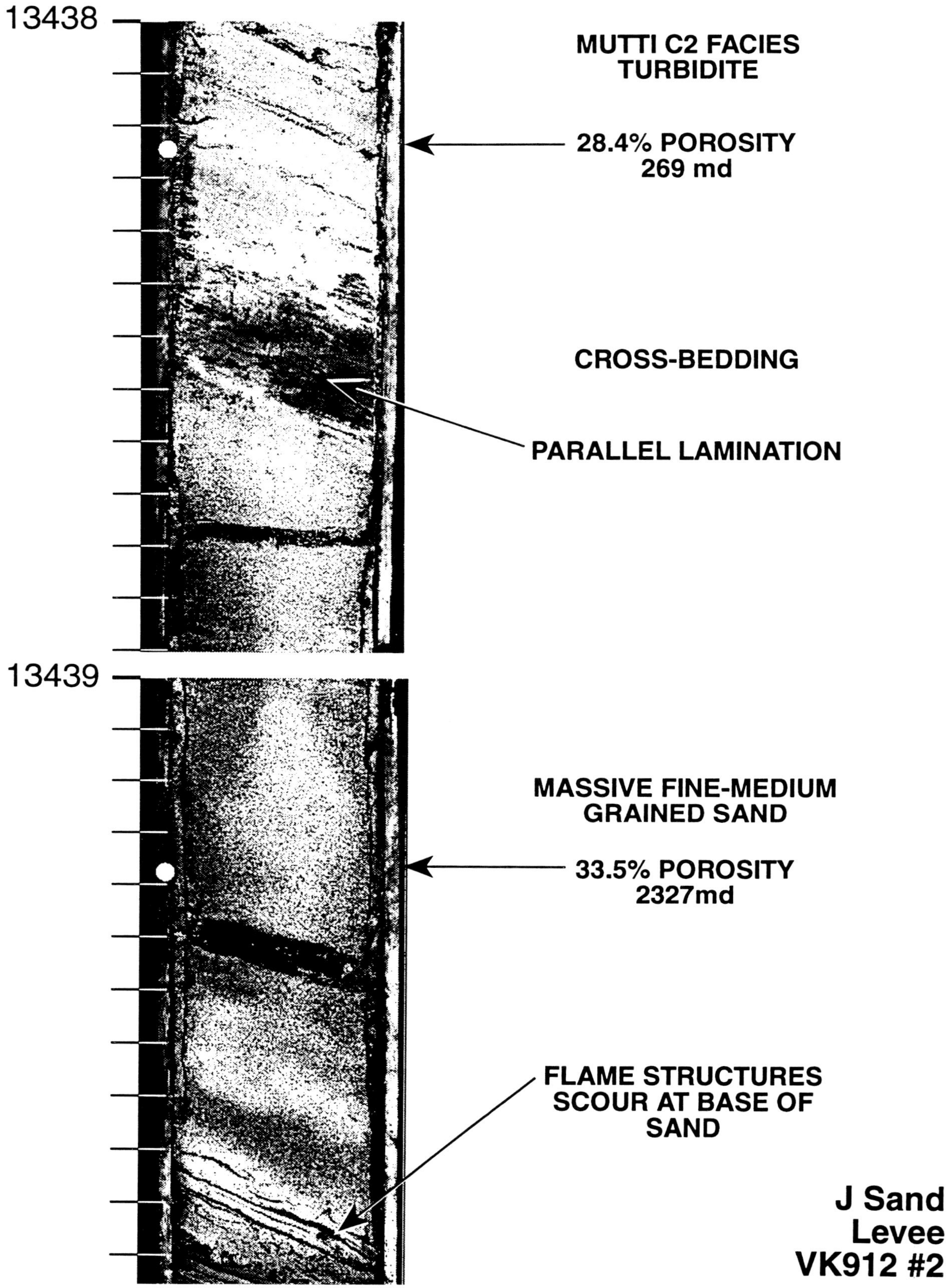

FIG. 17.—Core photograph of the J levee facies sands.

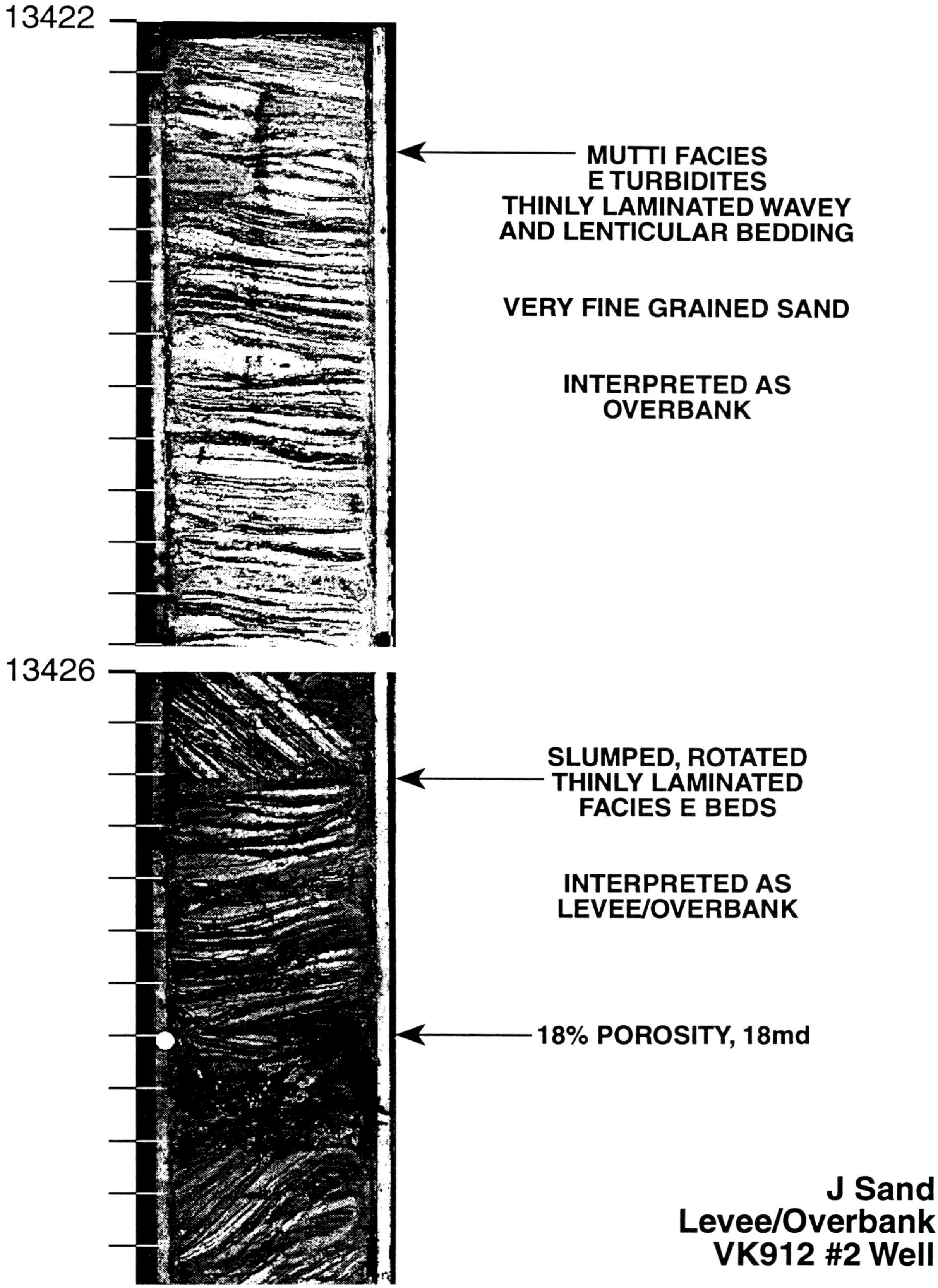

FIG. 18.—Core photograph of the J sand levee-overbank facies sands.

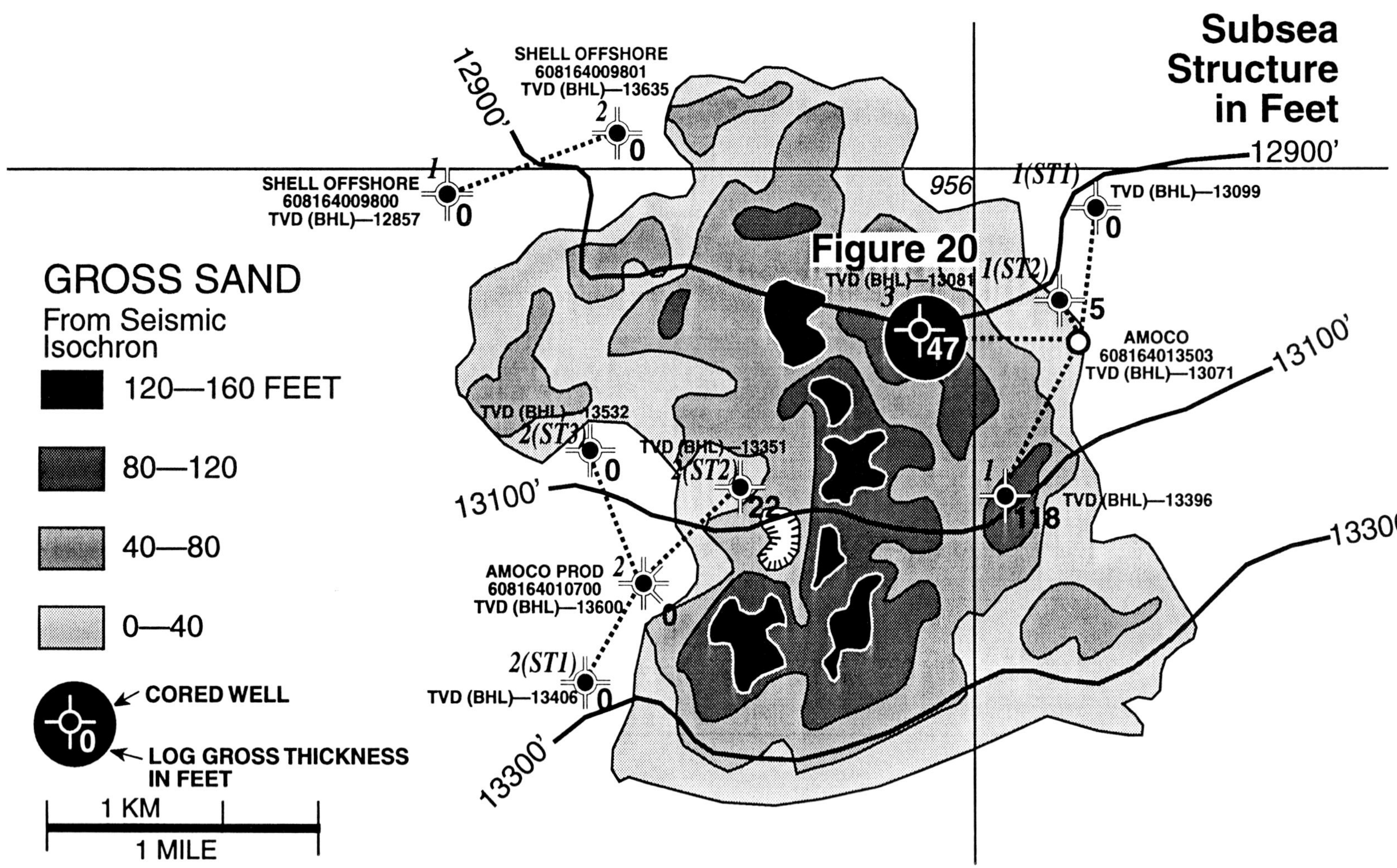

FIG. 19.—N fanlobe gross sand isopach with structural contours.

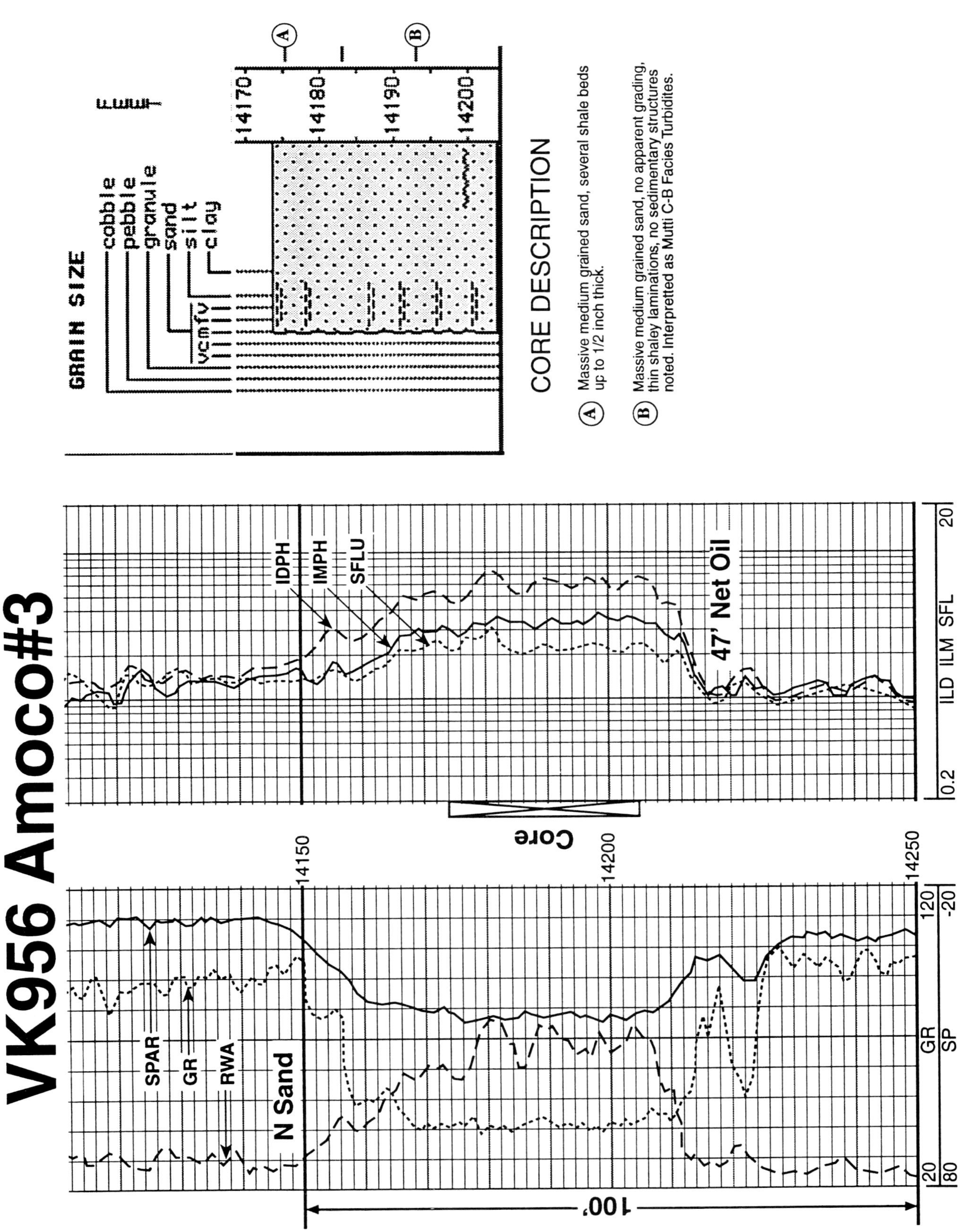

FIG. 20.—N sand inner fan channel facies. The cored interval is interpreted as a Mutti facies B turbidite.

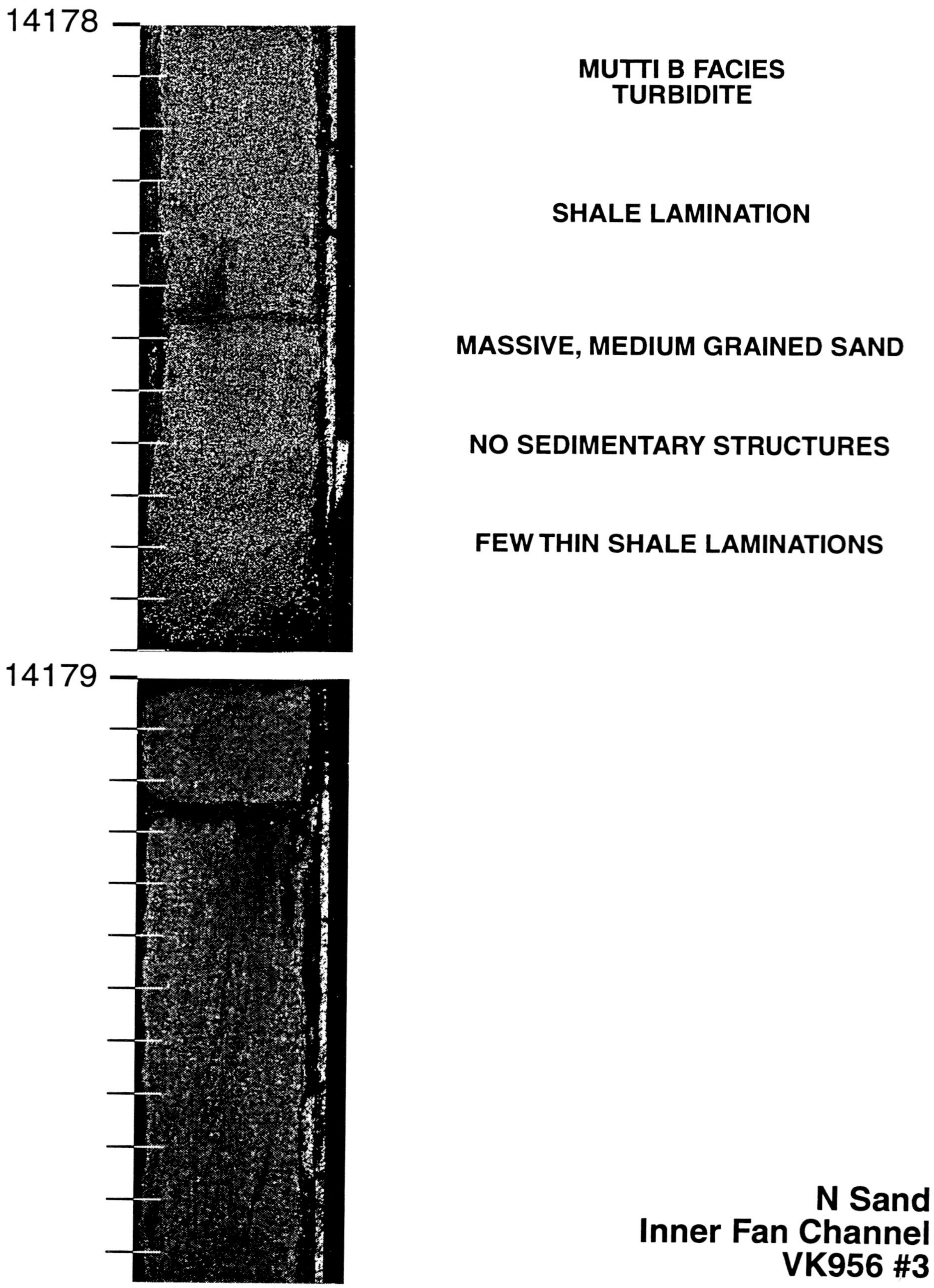

FIG. 21.—Core photograph of the N channel facies sands.

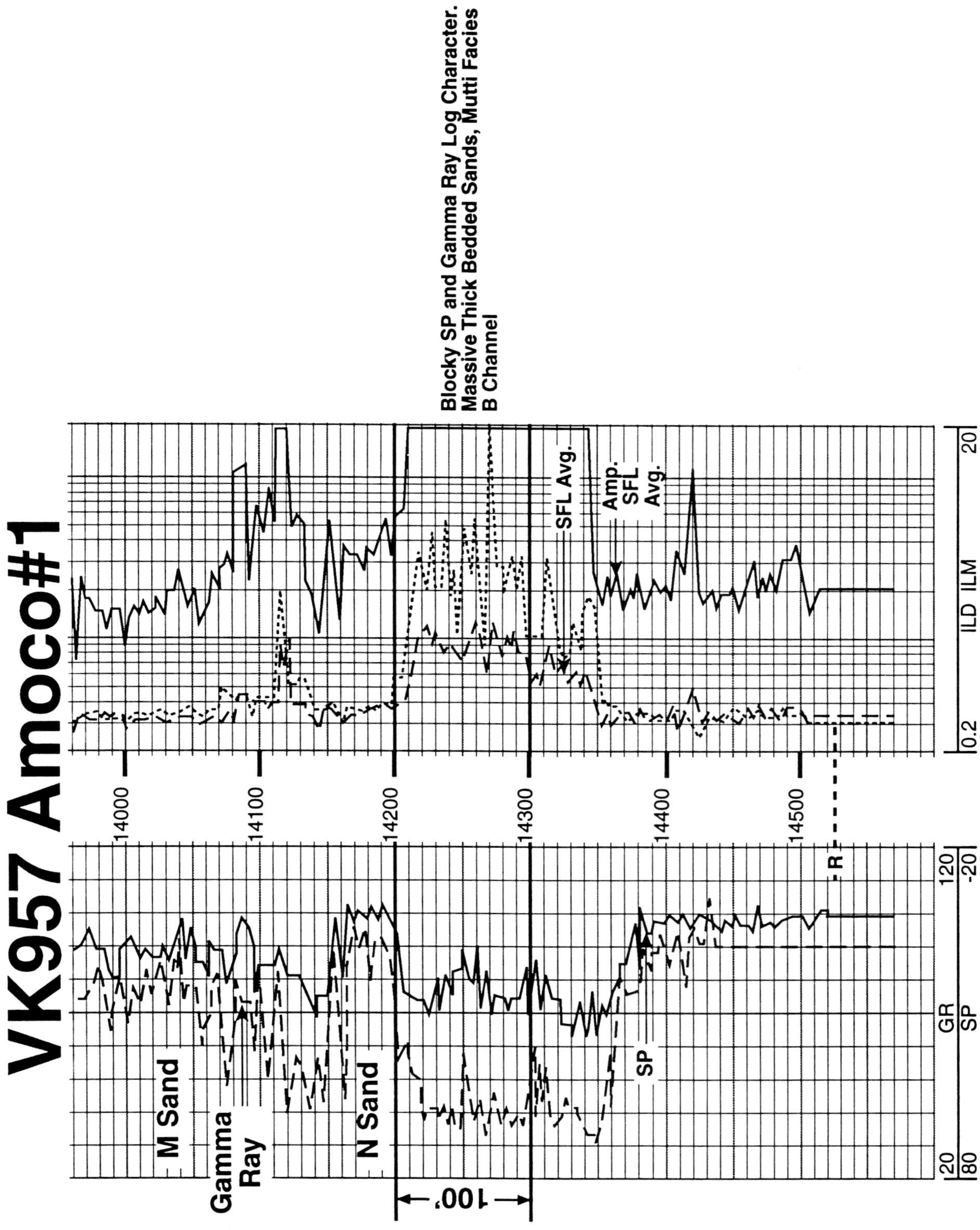

FIG. 22.—VK 957 #1 well log, N sand inner fan channel.

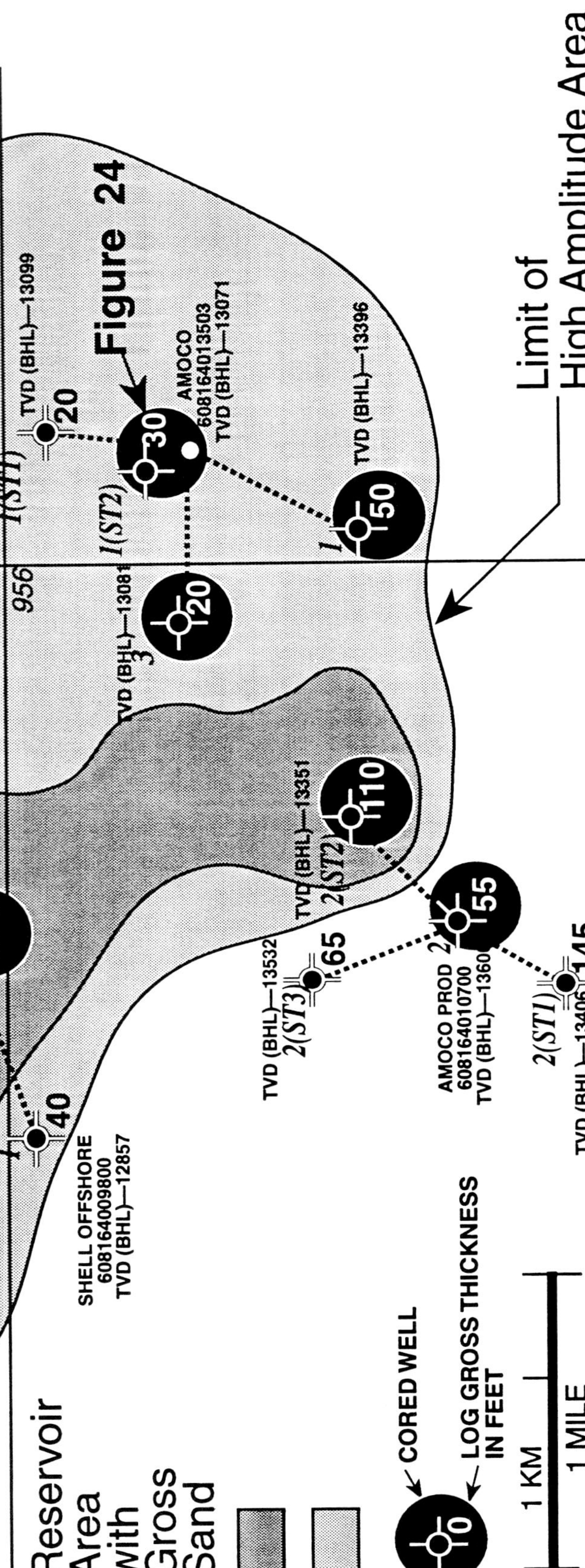

FIG. 23.—L fanlobe gross sand with reservoir outline.

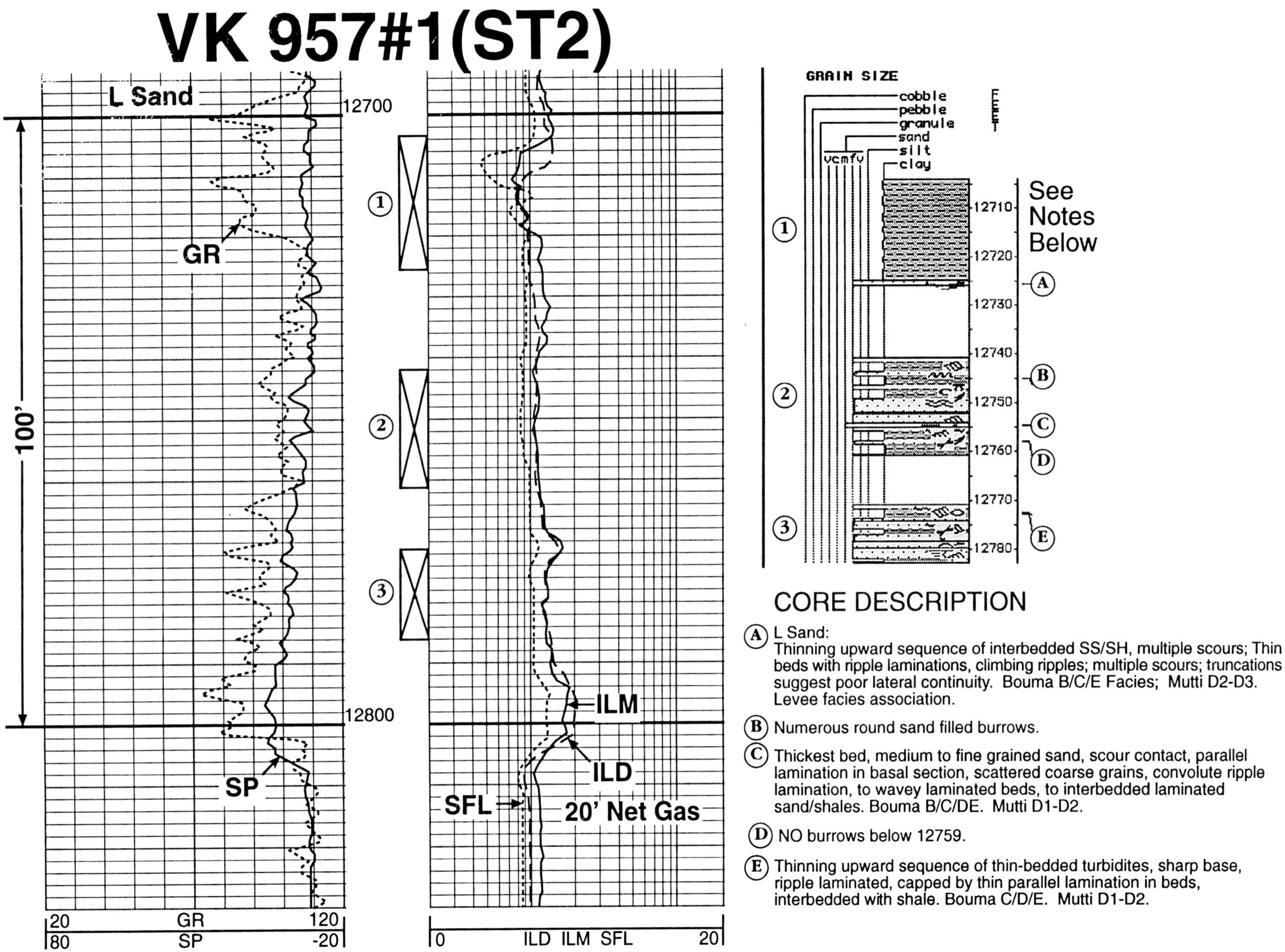

FIG. 24.—L sand levee-overbank facies. The cored intervals are interpreted as a Mutti facies D turbidites.

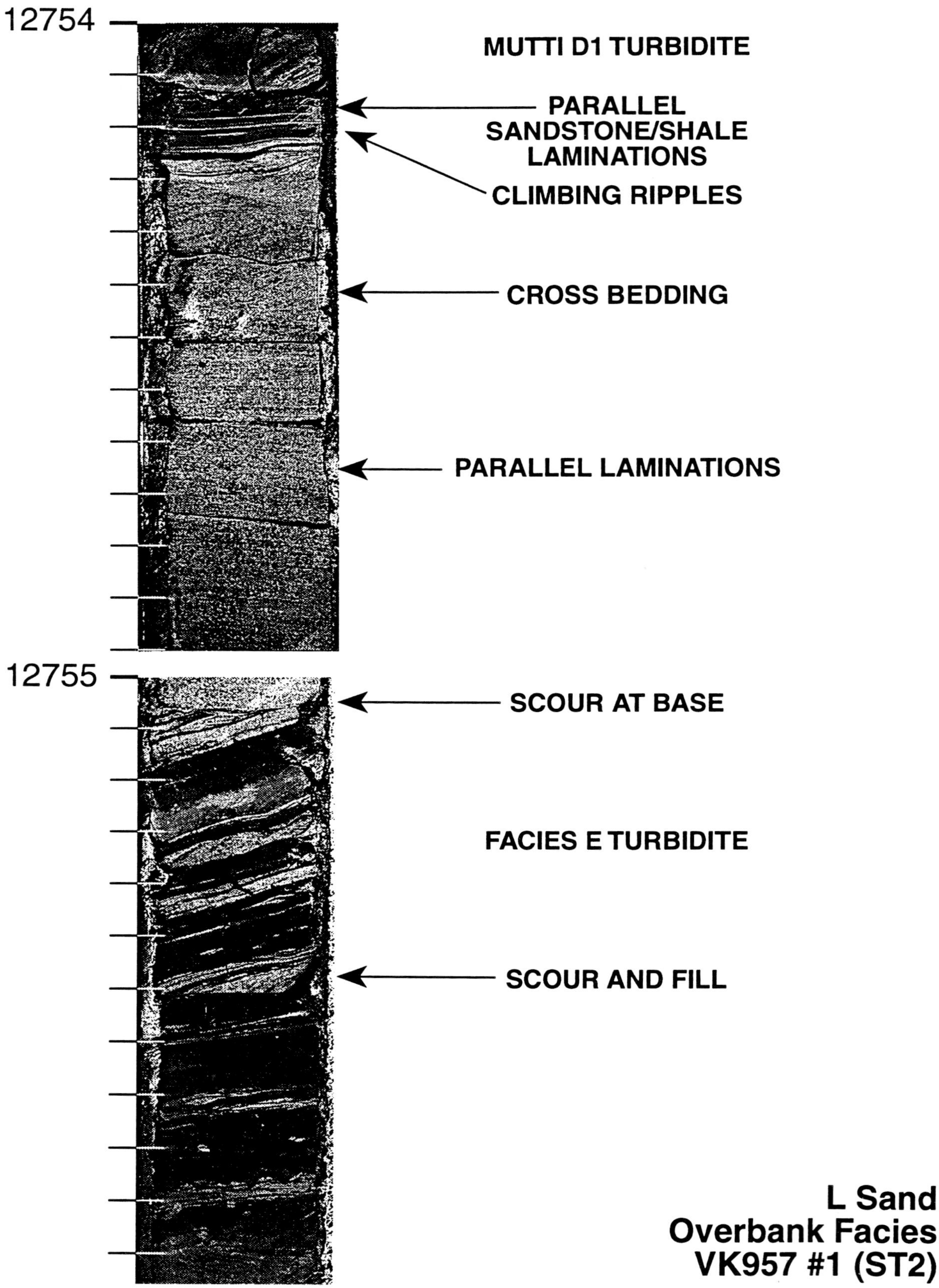

FIG. 25.—Core photograph of the L sand levee-overbank facies sands, Mutti D1 facies turbidite.

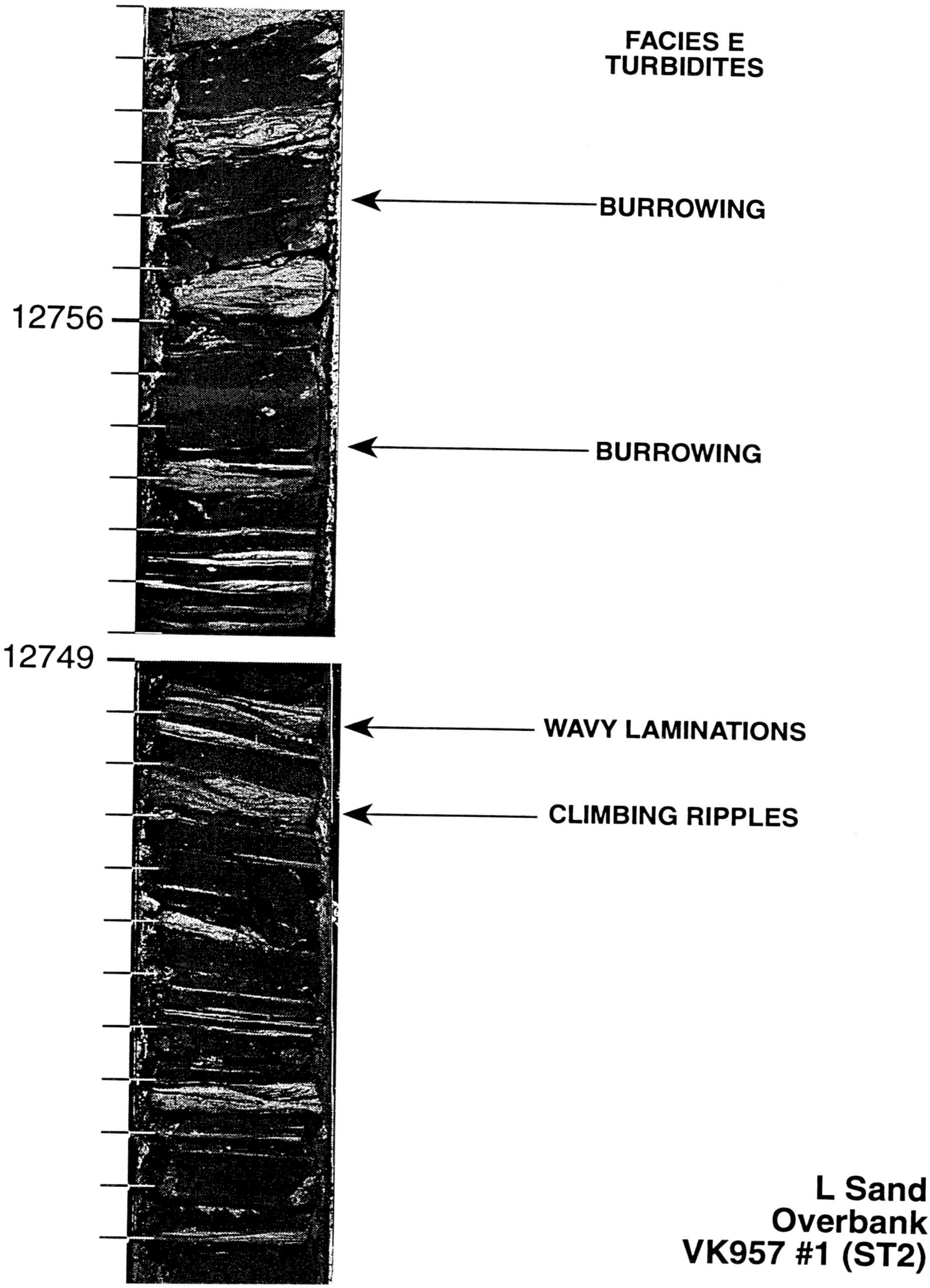

FIG. 26.—Core photograph of the L sand levee-overbank facies sands Mutti E facies turbidite.

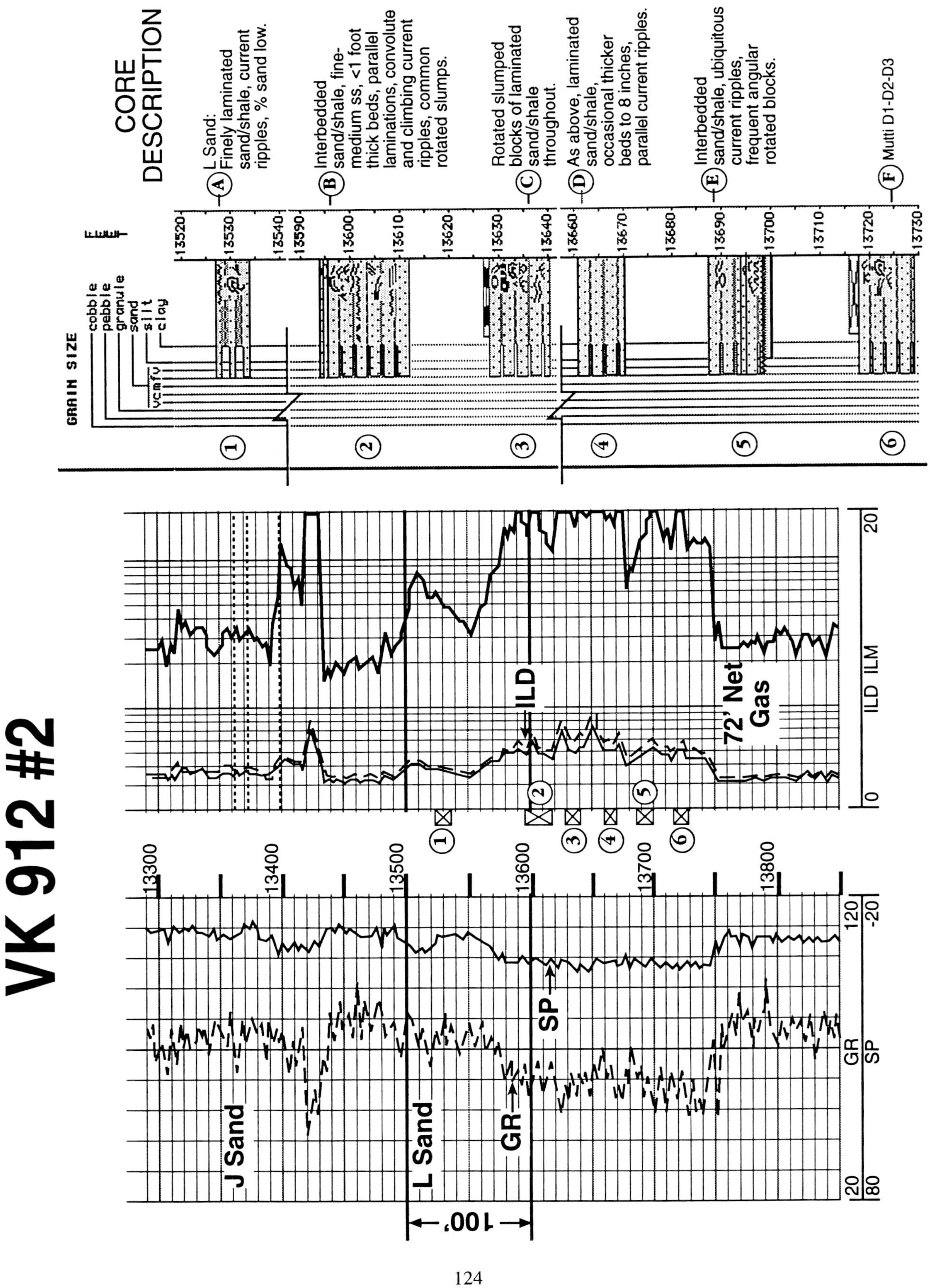

FIG. 27.—L sand levee-overbank facies. The cored intervals are interpreted as a Mutti facies D turbidites.

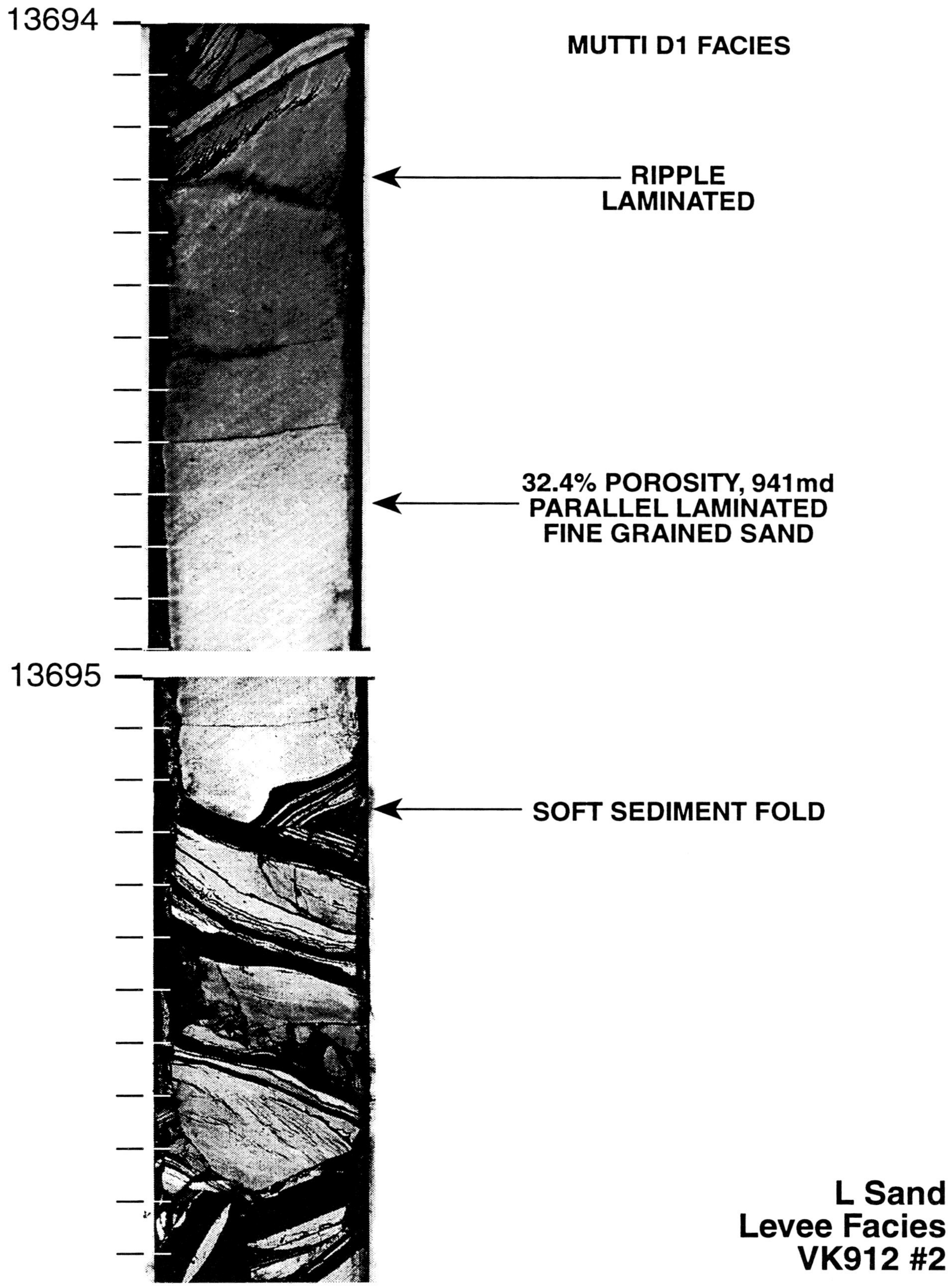

FIG. 28.—Core photograph of the L sand levee-overbank facies sands, Mutti D1 facies turbidite.

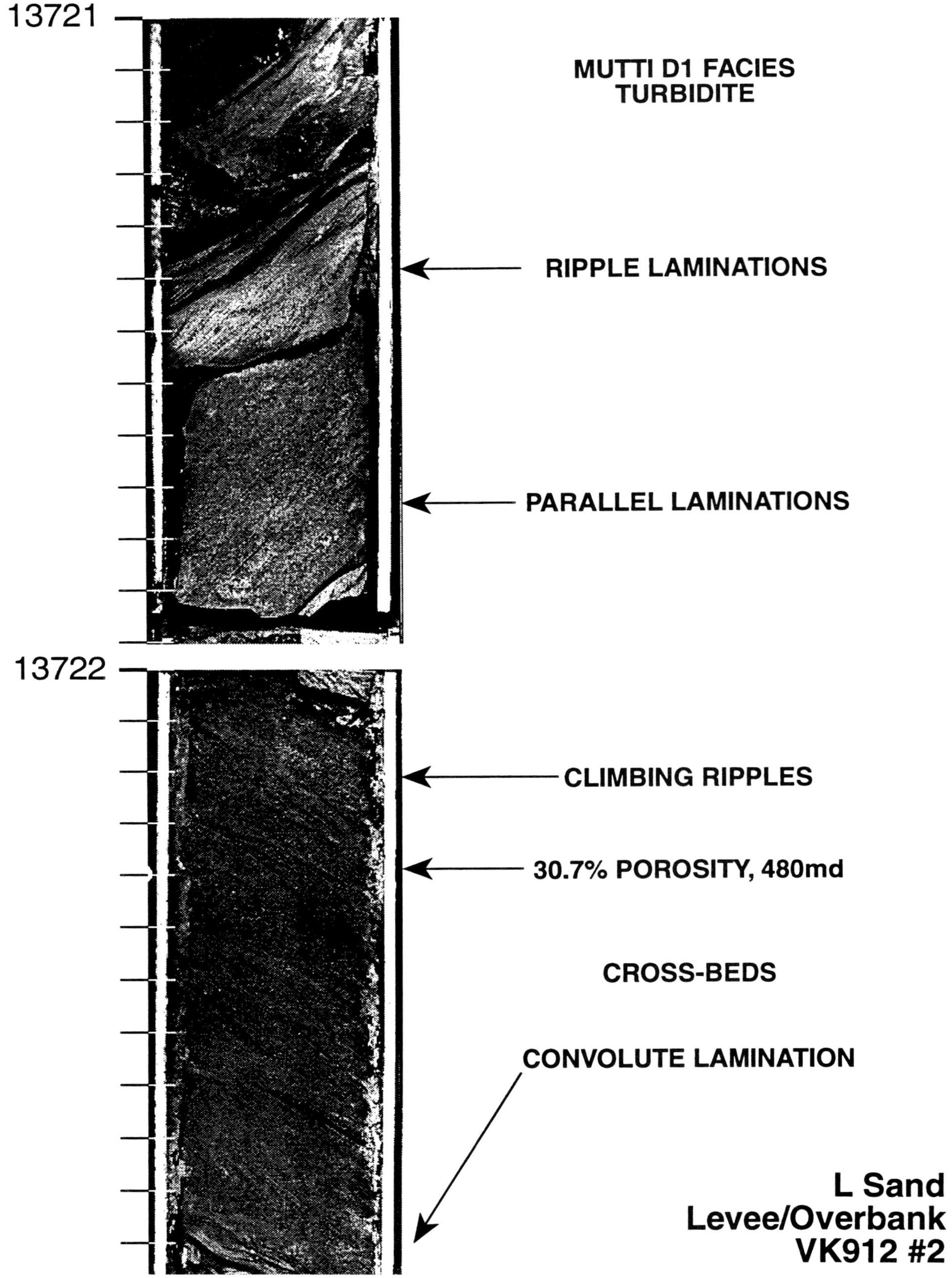

FIG. 29.—Core photograph of the L sand levee-overbank facies sands, Mutti D1 facies turbidite.

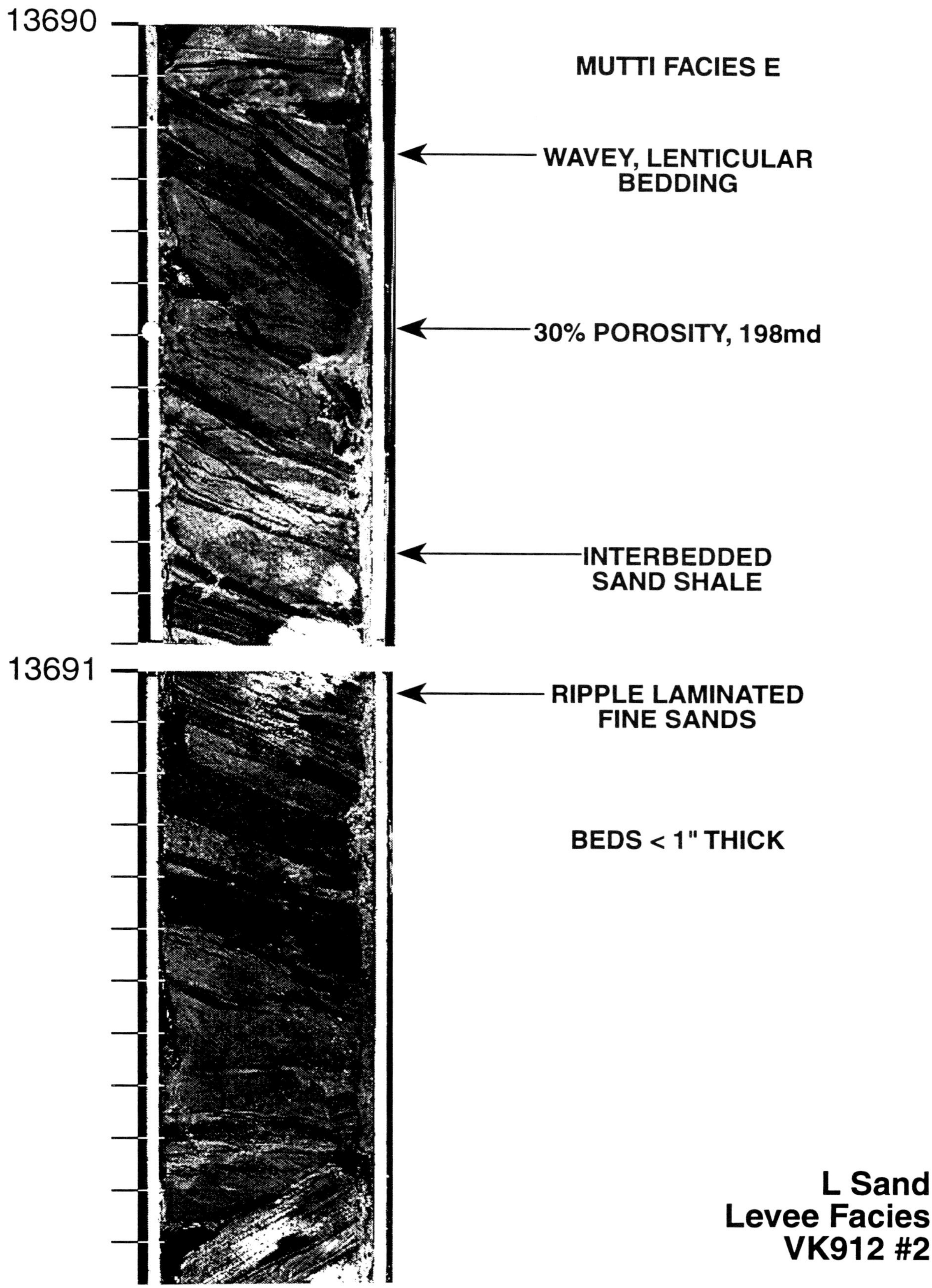

FIG. 30.—Core photograph of the L sand levee-overbank facies sands, Mutti E facies turbidite.

APPENDIX 1.

Core Description Legend

LITHOLOGY

SAND/SANDSTONE SILT/SILTSTONE SHALE/MUDSTONE breccia

CONTACTS

Scoured

PHYSICAL STRUCTURES

small current ripples	Climbing Ripples	Planar Tabular Bedding
High Angle Tabular Bedding	Wavy Parallel Bedding	Lenticular Bedding
Hummocky Cross-strat.	Convolute Bedding	Scour
Graded Bedding	Fault	Load Casts
soft sed slump	burrows	

LITHOLOGIC ACCESSORIES

Organic Shale Lamina Rip Up Clasts

REFERENCES

Haq, Hardenbol and Vail, 1987, in Atlas of Seismic Stratigraphy: AAPG Studies in Geology #27, v. 1.

Mutti, E., 1977, Distinctive thin-bedded turbidite facies and related depositional environments in the Eocene Hecho Group: Sedimentology, v. 24, p.107-131.

Mutti, E. and F. Ricci Lucchi, 1978, Turbidites of the Northern Apennines: Introduction to Facies Analysis: American Geological Institute Reprint Series #3, from International Geology Review, v. 20, #2, p. 125-126.

Nelson, H. C. and Tor H. Nilsen, 1984, Modern and Ancient Deep-sea Fan Sedimentation: SEPM. Short Course #14 Notes.

Reading, H. G. and M. Richards, 1994, Turbidite Systems in Deepwater Basin Margins Classified by Grain Size and Feeder System: AAPG Bulletin, v. 78, p. 792-822.

Sangree, J. B. and J. M. Widmier, 1977, Seismic interpretation of clastic depositional facies, in Seismic stratigraphy-applications to hydrocarbon exploration, AAPG memoir 26, Tulsa, Okla.

Shanmugam, G. and R. J. Moiola, 1991, Types of submarine fan lobes: models and implications: AAPG Bulletin, v. 75, #1, p. 156-179.

Stow, D., Howell, D., and Nelson, C. 1983, Sedimentary, tectonic, and sea level controls on submarine fans, in Normark, W. R., Bouma, A. H., eds, and Barnes, N. E., compiler, Submarine Clastic Systems; Deep-sea Fans and Related Facies: Geo-marine Letters, v. 3, p. 57-64.

Walker, R. G. ,1992, Turbidites and submarine fans, in Facies Models: response to sea level change, Geological Association of Canada.

WHOLE CORE OF MIDDLE PLEISTOCENE PAY ZONES IN GARDEN BANKS BLOCK 171, GULF OF MEXICO

ELIZABETH CERATES STRATHOUSE
Marathon Oil Company, 5555 San Felipe Road, Houston, TX 77056

ABSTRACT

Fifty-three feet of whole core were taken from the Marathon #2 well in Garden Banks Block 171, Offshore Louisiana, Gulf of Mexico to determine reservoir characteristics, detailed lithology and depositional environments of the middle Pleistocene *Trimosina A* zone. The *Trimosina A* zone was previously evaluated in the Marathon #1 well in Garden Banks Block 215 with wireline logs, sidewall cores and repeat formation tester pressures and samples. In the GB 215 #1 well, the *Trimosina A* contains 79 net ft of oil-filled, finely laminated, very fine-grained sand and silt interbedded with shale with good reservoir pressure, an average porosity of 27% and estimated permeabilities of 50-1120 millidarcies.

In the GB 171 #2 well, conventional Cores 1 and 2 were taken below a 20 ft oil sand and contain mostly shale, but also unexpected amounts of poorly lithified, finely laminated silt filled with oil. Core 3 was taken near the base of the thickest oil sand (40 ft thick) in the well and contains unconsolidated, interlaminated, well-sorted, subangular to well-rounded silty sand and massive, medium-grained, angular to well-rounded, moderately to moderately-well sorted sand. The average porosity of the oil sand in Core 3 is 27.9% and the geometric mean of the core plug permeability measured under 2,000 psi confining pressure is 671 millidarcies.

The amount of dip in the cores was considerably higher than that of seismically visible horizons. The cores preserve irregular and anomalously high dips ranging from 50 to 60 degrees as well as gentler dips of 10 to 20 degrees. Images as well as structural attitudes were evaluated from the Formation Microscanner log (FMS) and confirm those observed in the core. In addition to confirming dip measurements in the cores, the FMS images show detailed dip patterns that were not evident from the dipmeter. While the dips are irregular and anomalously high, dips are consistent over 5 to 20 ft intervals. Furthermore, the cores show deformation features such as fine-scale folds, soft-sediment shearing and non-horizontal, closely-spaced fractures in Cores 1 and 2. Core 3 shows contorted laminations. The nature of the dips and the structures in the core indicate that the *Trimosina A* zone has been deformed into small, rotated blocks. This results in compartmentalizing the hydrocarbons which was corroborated by the drill-stem test results. Primary sedimentation structures in all three cores are too deformed to suggest much about depositional environments.

In summary, whole core from the Marathon #2 well in Garden Banks Block 171, Offshore Louisiana, Gulf of Mexico, was crucial to understanding poor reservoir potential from good reservoir quality sand and economically disappointing hydrocarbon recoveries from drill-stem tests of the middle Pleistocene *Trimosina A* zone.

INTRODUCTION

The Garden Banks Block 171 area is located approximately 190 miles off the coast of Louisiana in the Gulf of Mexico in 1,000 ft of water (Fig. 1). The stratigraphy in this area has been subdivided into lower and upper Pliocene and lower, middle and upper Pleistocene according to benthic foraminifera occurrences after Steinmetz (1986). The structure at Garden Banks Blocks 171 is complicated (Figs. 2 and 3). The structure in the middle Pleistocene

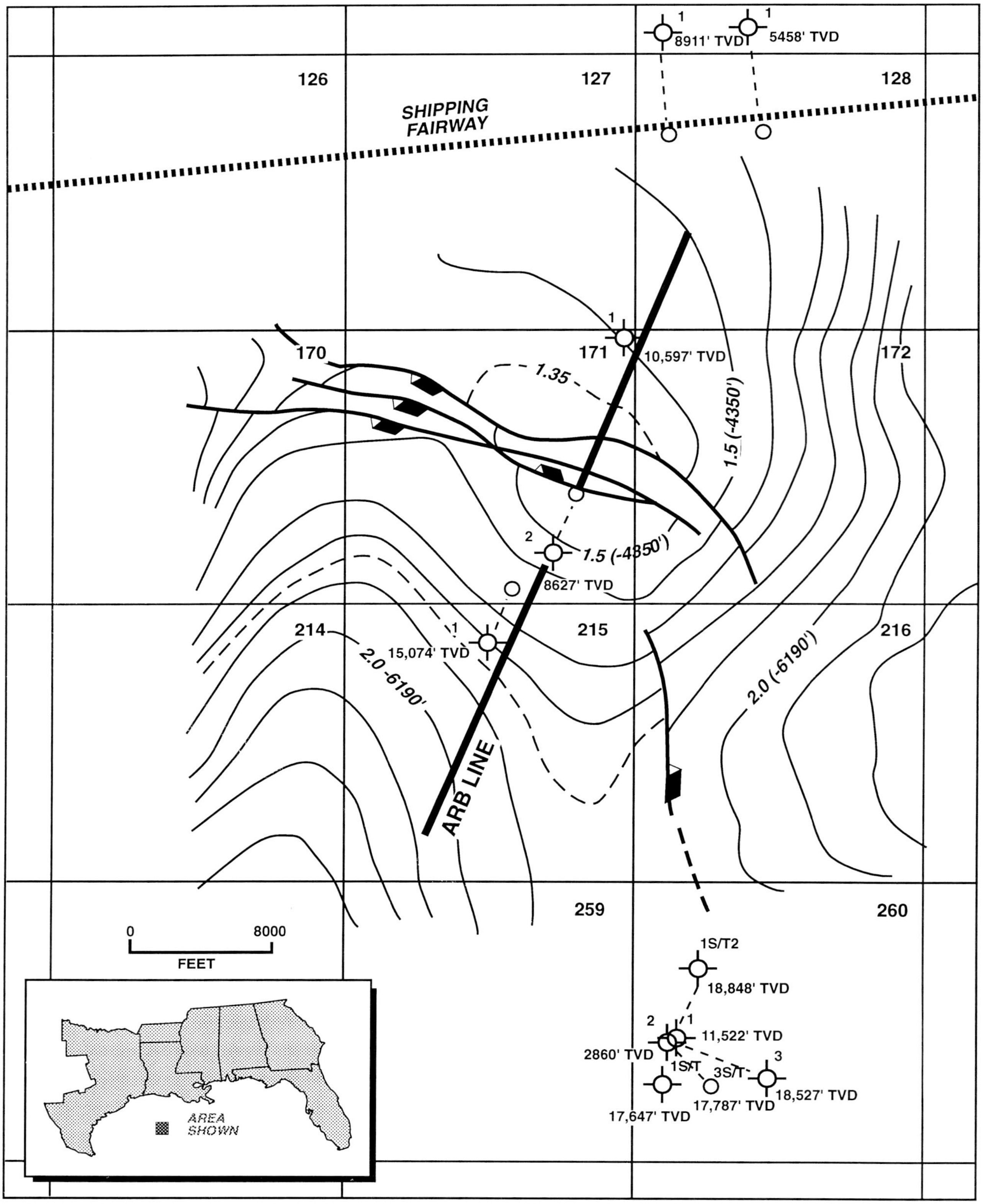

FIG. 1.—Index map of the Garden Banks Blocks 171, 172 and 215 prospect area showing time structure on top of the *Trimosina A* zone. Contour interval=100 milliseconds, 50 millisecond contour dashed. Scale in=8,000 ft. TVD=true vertical depth. See Figures 2 and 3 for seismic reflection line ARB LINE.

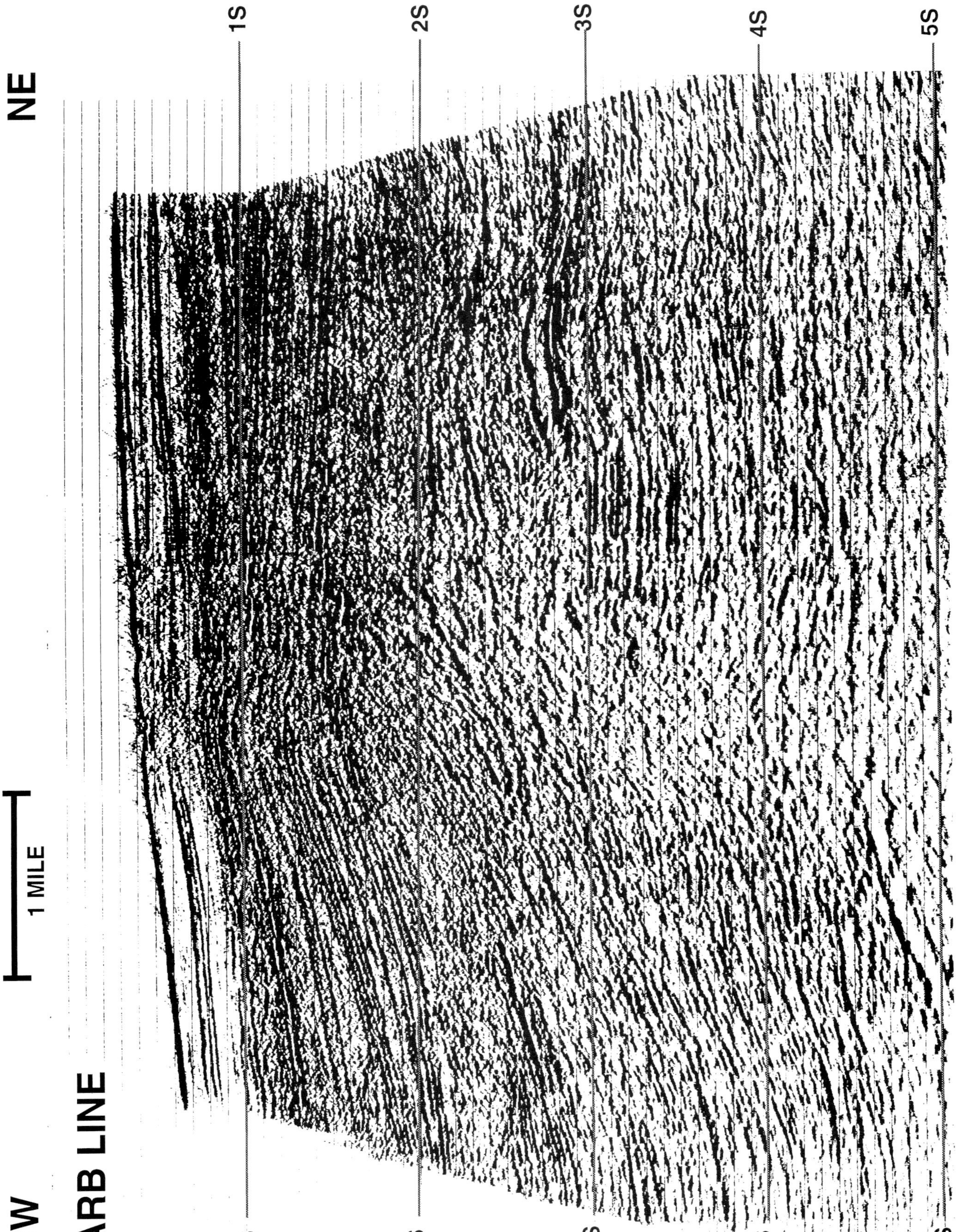

FIG. 2.—Uninterpreted seismic section ARB LINE. See Figure 1 for location. Vertical scale in seconds.

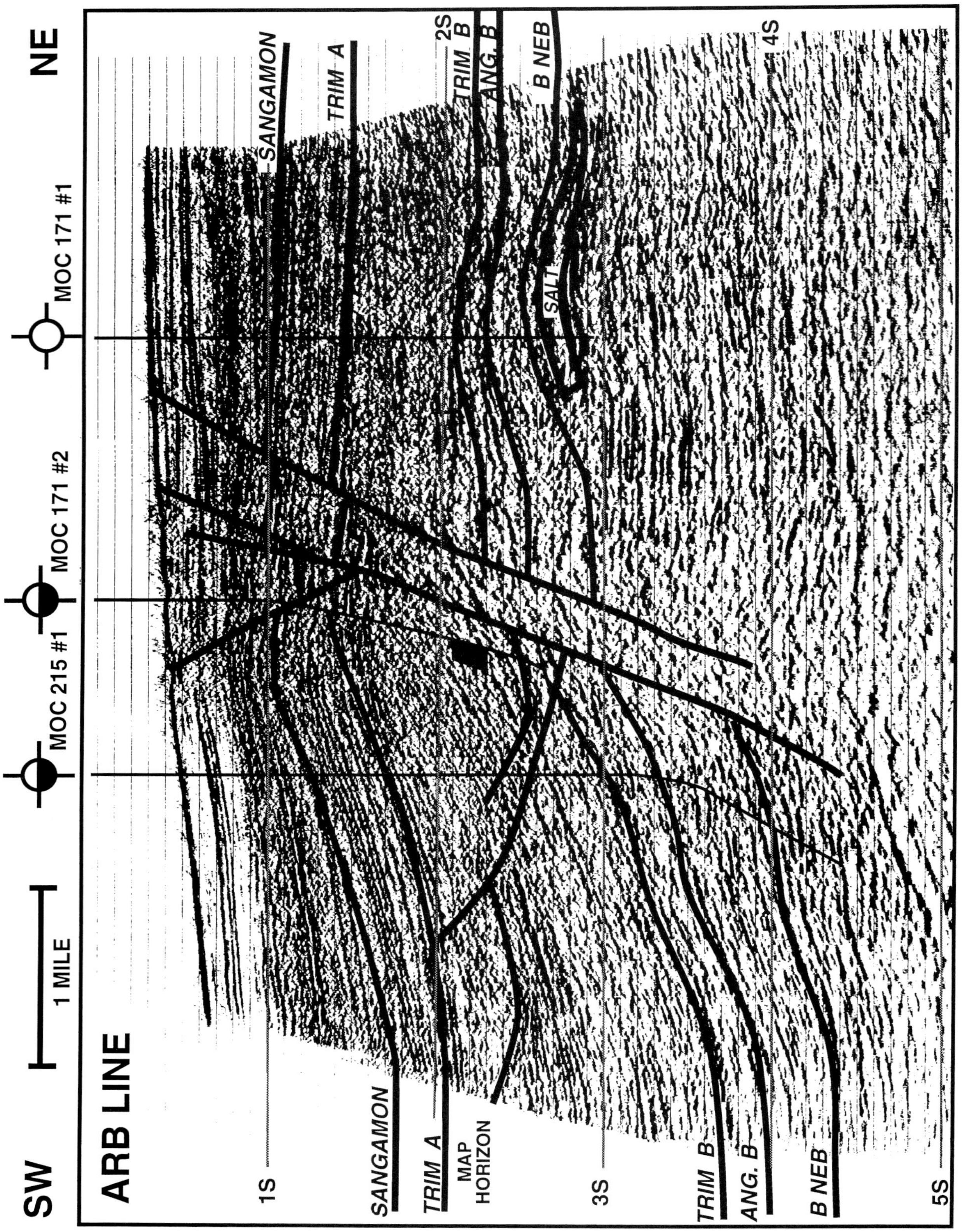

FIG. 3.—Interpreted seismic section ARB LINE showing well locations, faults, salt and biostratigraphic horizons. Vertical scale in seconds. Core locations (Figs. 4 and 5) are indicated along the GB 171 #2 wellbore.

(*Sangamon* and *Trimosina A*) is a faulted anticline cut by northwest-trending growth faults. An interval of cut and fill has been interpreted in the discontinuous seismic facies just below the *Trimosina A* horizon (Figs. 2 and 3). The middle Pleistocene to upper Pliocene (*Trimosina B* to *Basal Nebraskan*) stratigraphy is expanded in the downthrown fault block. On the upthrown side of the fault, a salt tongue occurs within the upper Pliocene age (*Basal Nebraskan*) section. Deeper salt may underlie the large anticlinal structure (below 5 seconds two-way travel time) deeper than the section shown on ARB LINE (Figs. 2 and 3).

Three different objectives have been drilled in the Garden Banks (GB) Block 171 area (Fig. 1, 2 and 3). Marathon and Exxon drilled the GB 171 #1 in 1984 through a salt tongue and encountered lower bathyal sediment of late Pliocene age below salt. Marathon and partners Amerada Hess, Phillips and TXP drilled GB 215 #1 four years later in 1988, into the deeper, downthrown rollover. The well encountered 79 net feet of oil pay in upper to mid-bathyal, middle Pleistocene *Trimosina A*, as well as oil and gas shows in lower bathyal, late Pliocene age sand. The *Trimosina A* pay zone in the GB 215 #1 was evaluated with wireline logs, sidewall cores and repeat formation tester (RFT) pressures and samples. Based on analyses of the data, the pay zone contains 79 net ft of oil-saturated, finely laminated, very fine-grained sand and silt interbedded with shale. The RFT recovered oil. The pay zone has average porosity of 27% and estimated permeability of 50-1120 millidarcies from sidewall core analyses. GB 171 #2, from which whole core was taken, was drilled by the same partners the next year to test the updip equivalent to middle Pleistocene *Trimosina A* zone oil sand and wet *Trimosina B* zone sand encountered in the GB 215 #1 well, near the top of an anticlinal closure. The GB 171 #2 bottomed just above the top of the *Trimosina B* based on seismic and stratigraphic ties. Hydrocarbons fill every pore space in the numerous sands and silts encountered in the *Trimosina A* objective. Whole core was taken from the GB 171 #2 well to determine reservoir characteristics, detailed lithology and depositional environments of the middle Pleistocene *Trimosina A* zone.

Amerada Hess discovered oil six miles south of Garden Banks Block 171 in GB 260. Reserves have been estimated at 200 million barrels of oil equivalent from lower Pleistocene and Pliocene sands. The nearest middle Pleistocene pay was found by Mobil at Garden Banks Block 72, 34 miles north-northwest. Their three wells averaged 110 net feet of hydrocarbons in middle Pleistocene sands.

METHODS

Core descriptions, wireline logs, tests and show information from the GB 171 #2 well are represented by schematic diagrams in Figures 4 and 5. Wireline logs run in the well include resistivity, spontaneous potential, sonic, gamma ray, neutron-density, formation microscanner, repeat formation tester and sidewall cores. Both images and structural attitudes were recorded with the formation microscanner. Sidewall cores were analyzed for porosity, permeability, saturations and lithology.

Whole core was taken by 30 ft increments and frozen in aluminum sleeves at the well site. The cores were shipped frozen to the core laboratory and were examined by x-ray fluorescence to determine maximum dip direction. Once slabbed, selected intervals were double plugged and analyzed after being photographed. The core plug analysis included porosity and permeability measured under confining pressure, saturations, lithology descriptions, X-ray diffraction, scanning electron microscope and thin section analysis. Additionally, special core analysis was performed on duplicate core plugs to analyze formation resistivity factor, formation resistivity index, oil-brine capillary pressure, gas-oil and water-oil relative permeability and pore volume compressibility.

CORE DESCRIPTION

Three 30 ft cores were taken from two zones and a total of 53 ft were recovered (Figs. 4 and 5). All of the sand, silt and silty shales in Cores 1, 2 and 3 have bright, light-yellow to white fluorescence, indicating the presence of hydrocarbon saturation which can be noted in Figures 6-11. The photos exposed to ultraviolet light are shown on the right of the photos exposed to plane light. Coring-induced fractures, which are probable zones of core rotation, are generally perpendicular to the core barrel and are typically injected with reddish drilling mud which make them readily identifiable.

Cores 1 and 2 From First Zone

The sediments of Cores 1 and 2 are made up of laminated silt, shaley silt and shale (Figs. 6 and 7). The sediments are very poorly lithified. The laminated silt is in

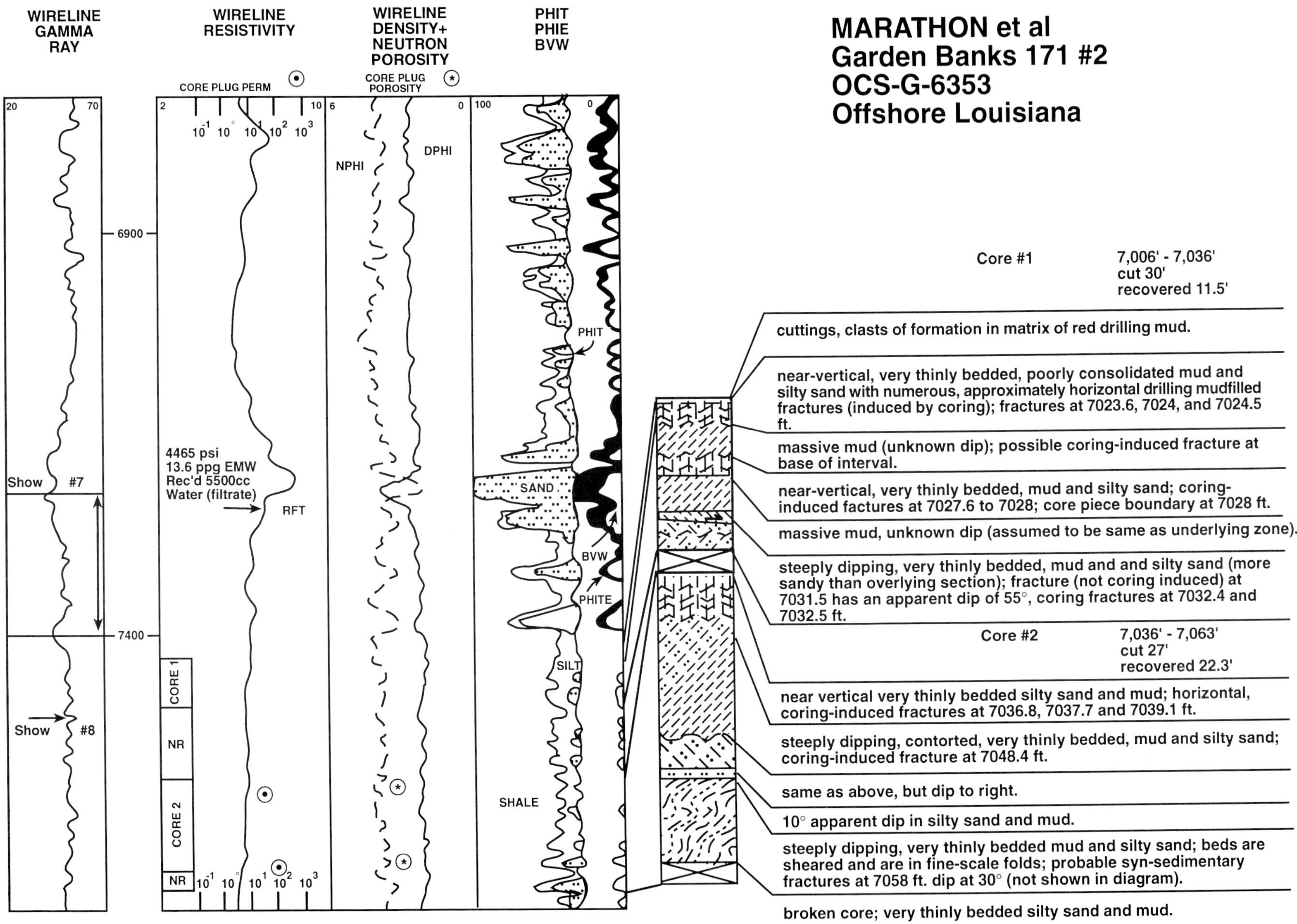

FIG. 4.—Wireline logs, core plug permeability and porosity (measured under 2,000 psi pressure), core description, shows and RFT results for core 1 and 2. PHIT=total porosity, PHIE=total effective porosity (excludes bound water in clays), BVW=bulk volume water, calculated from wireline logs.

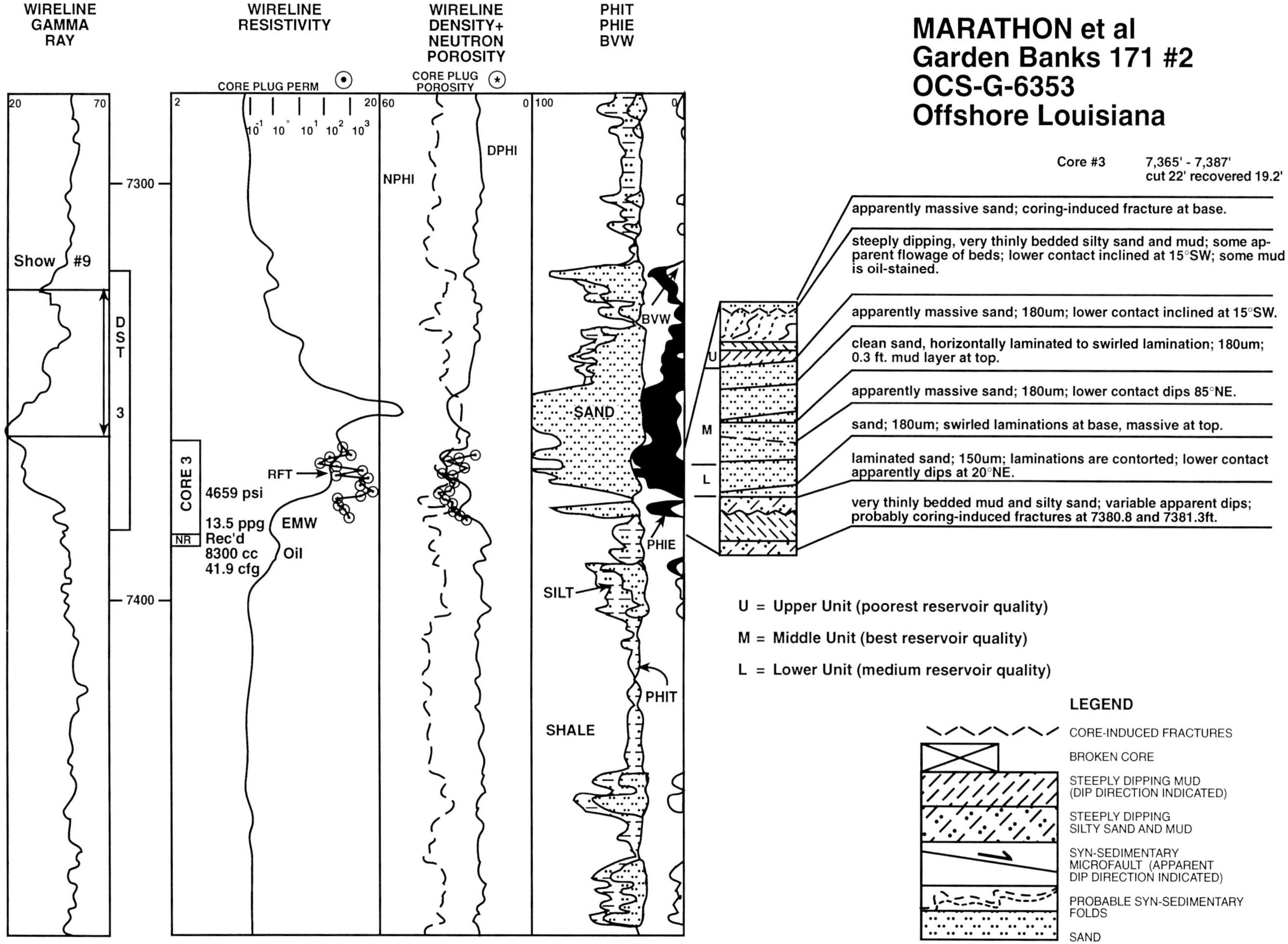

FIG. 5.—Wireline logs, core plug permeability and porosity (measured under 2,000 psi pressure), core description, shows and RFT results for core 3. PHIT=total porosity, PHIE=total effective porosity (excludes bound water in clays), BVW=bulk volume water, calculated from wireline logs.

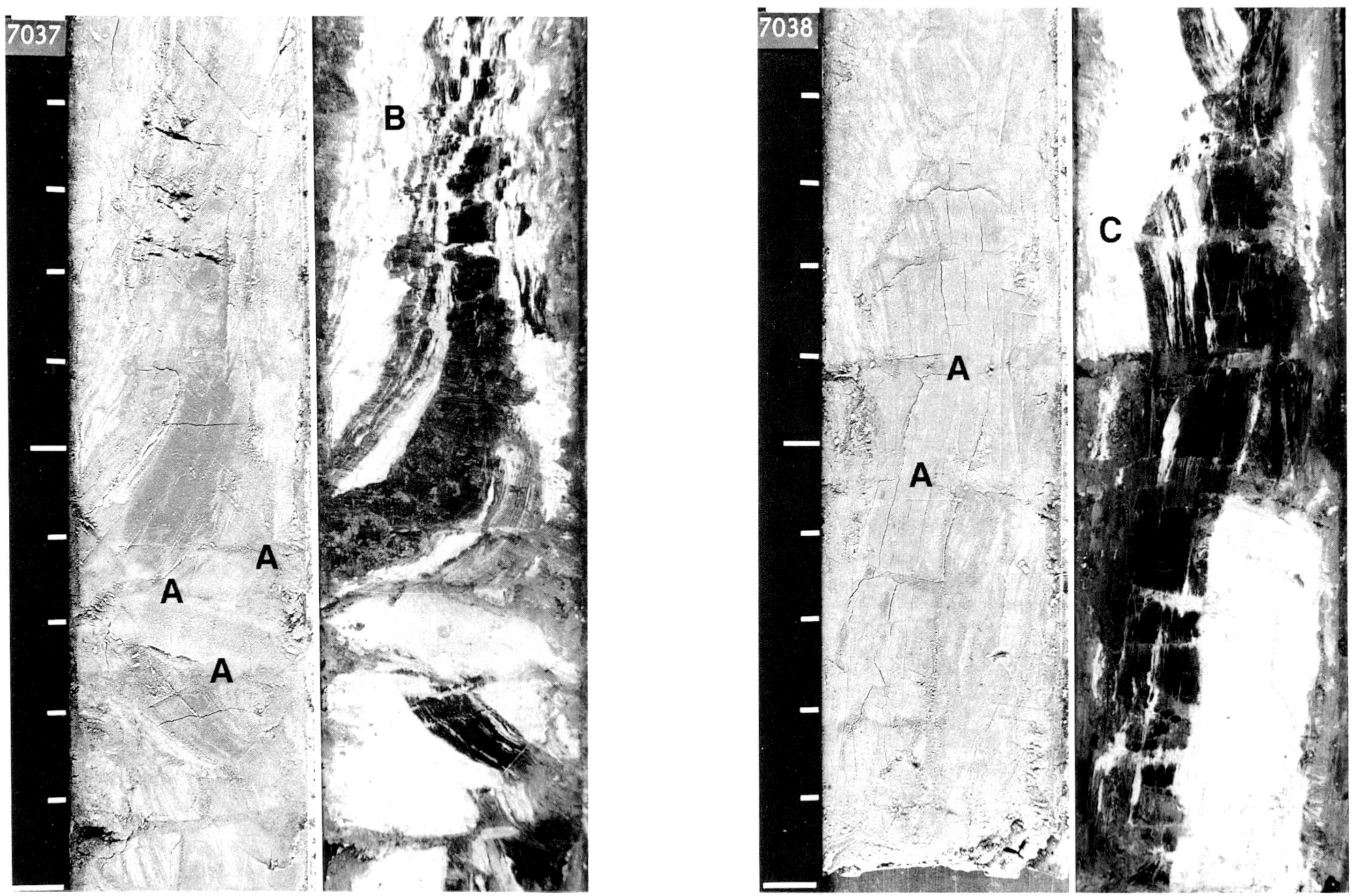

FIG. 6.—Whole core 7,037-7,039 ft MD showing coring induced fractures (A), nearly vertical dip, sheared beds of laminated silt (B) and small fractures cutting laminated silt and shale (C). Plane light on left, black light on right. Oil saturated silt and shaley silt is light colored and shale is dark colored on black light photos. Scale 5 in=1 ft. Footage marked in tenths.

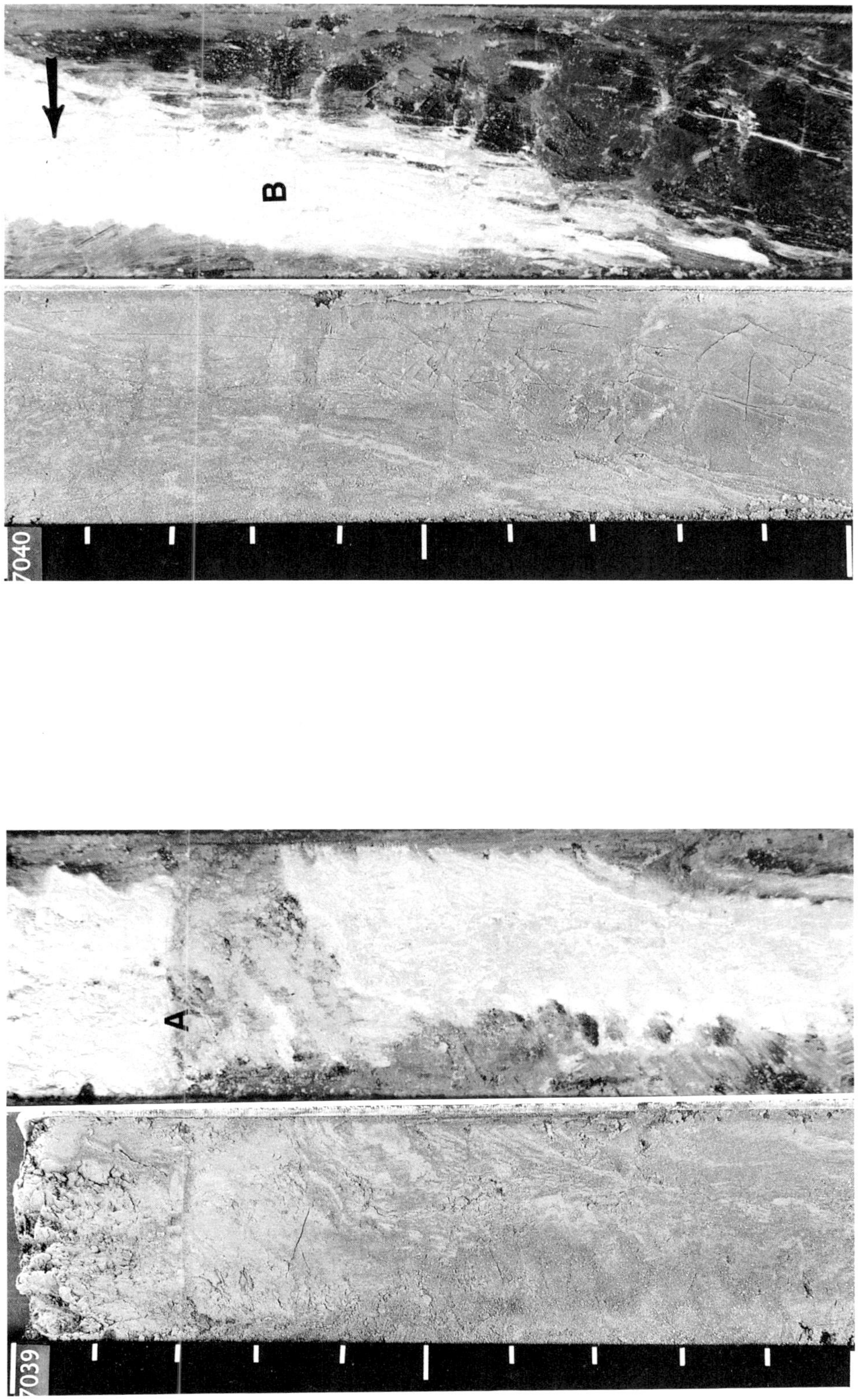

FIG. 7.—Whole core 7,039-7,041 ft MD showing coring induced fractures (A), nearly vertical dip, sheared beds (B) and small fractures. Plane light on left, black light on right. Oil sataurated silt and shaley silt is light colored and shale is dark colored on black light photos. Arrow=plugs taken. Scale 5 in=1 ft. Footage marked in tenths.

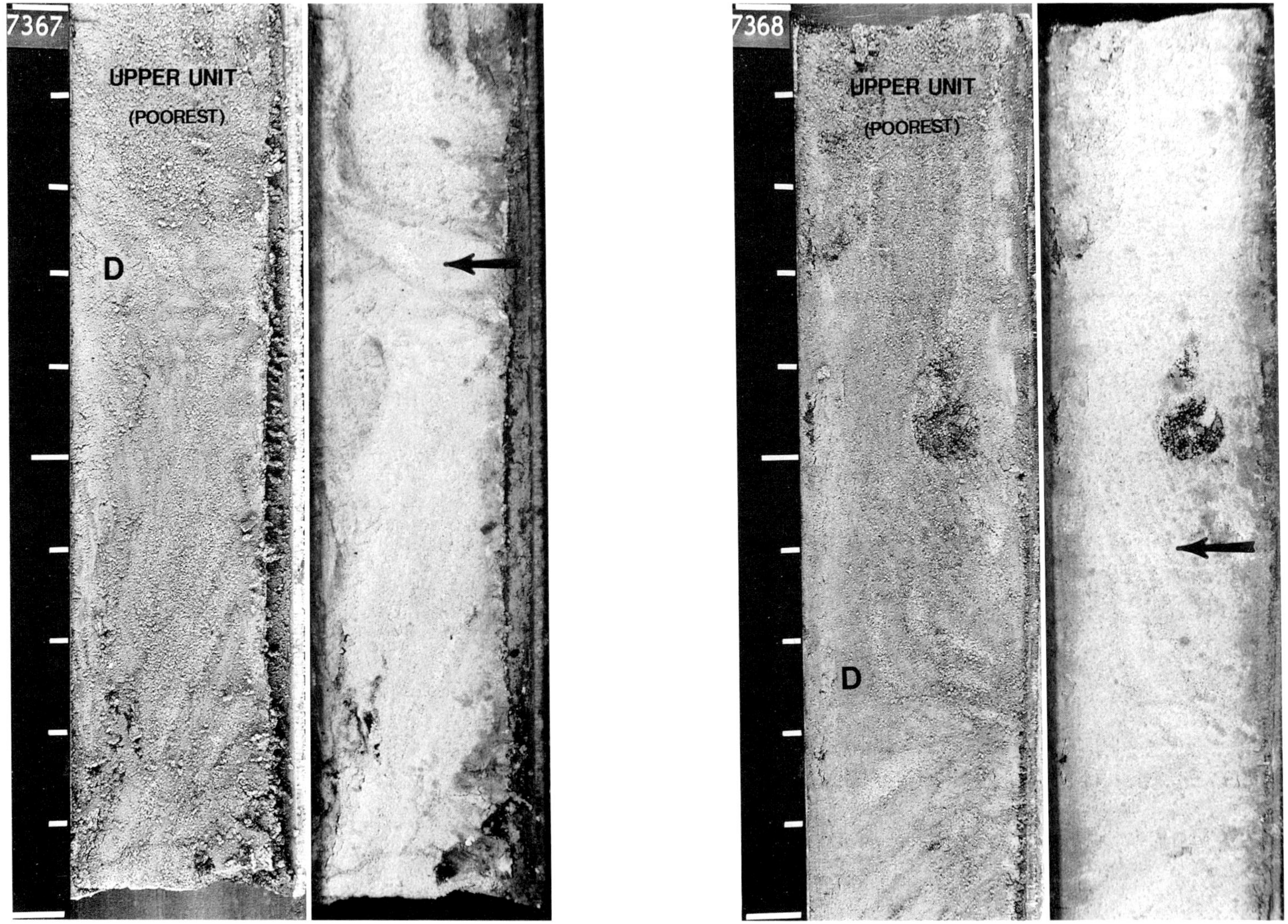

FIG. 8.—Whole core 7,367-7,369 ft MD showing contorted bedding and laminations (D). Arrow=plugs taken. Upper flow unit of core 3 (see Fig. 13). Plane light on left, black light on right. Scale 5 1/2 in=1 ft. Footage marked in tenths.

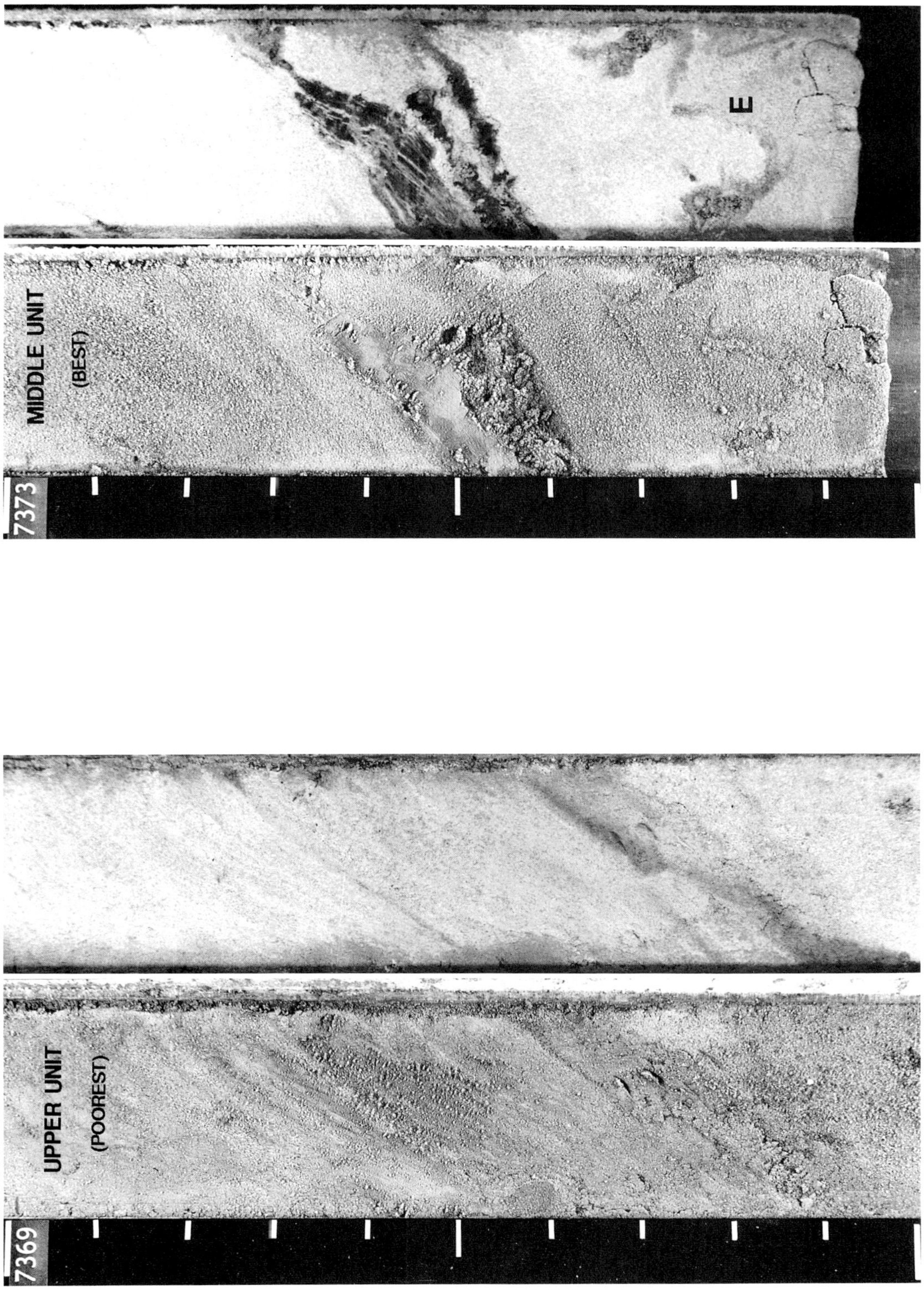

FIG. 9.—Whole core 7,369 ft MD and 7,373 ft MD showing contorted bedding and laminantions. Upper and middle flow unit of core 3 (see Fig. 13). Plane light on left, black light on right. White is oil saturated sand, grey is silty sand and black is shale on black light photo. Scale 5 1/2 in=1 ft. Footage marked in tenths. E=irregular bedding contact probably from soft sediment deformation.

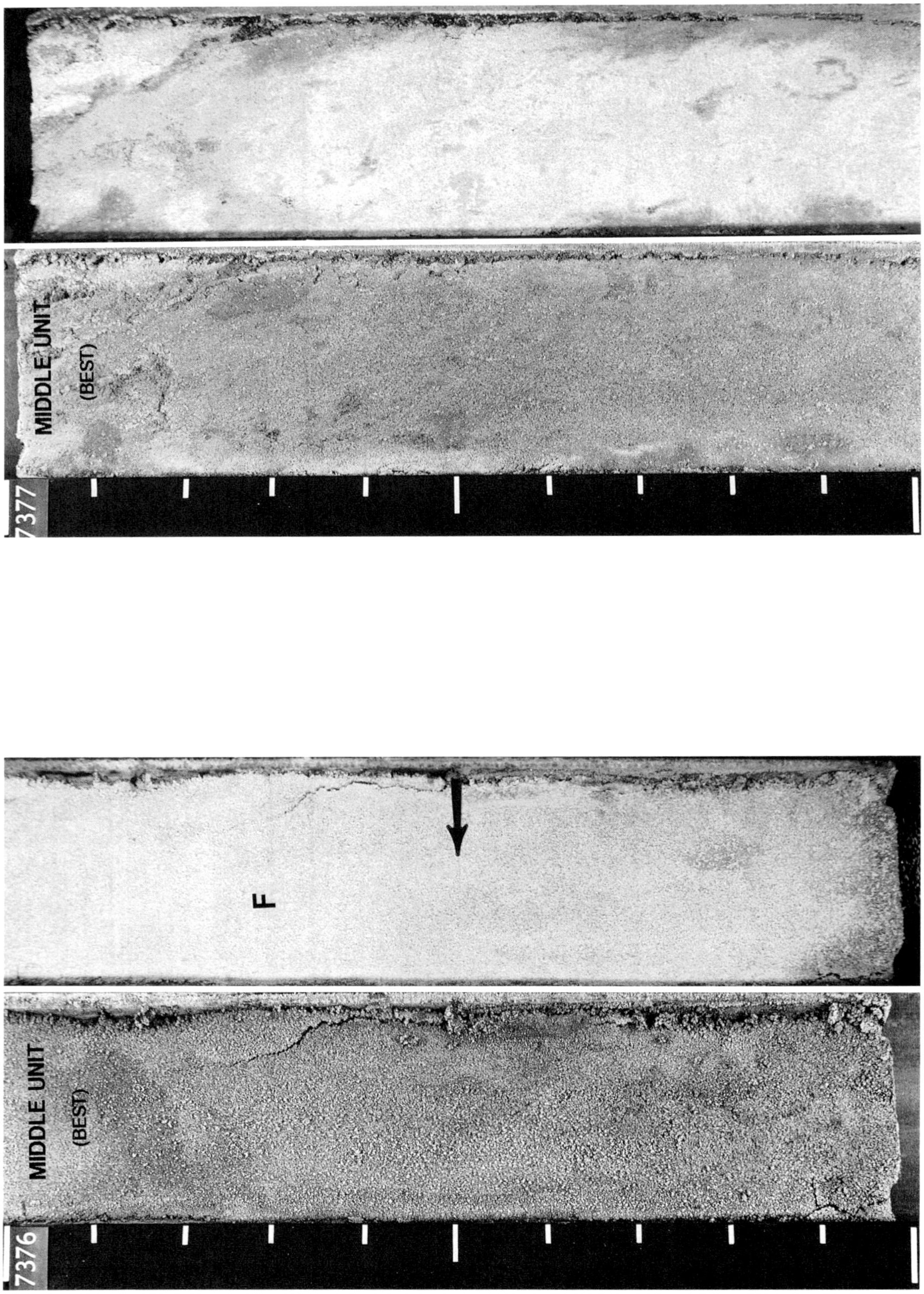

FIG. 10.—Whole core 7,376 ft MD showing structureless sand (F). Whole core 7,377 ft MD showing mottled sand, both of the middle flow unit of core 3. Arrow=plugs taken. Plane light on left, black light on right. Scale 5 1/2 in=1 ft. Footage marked in tenths.

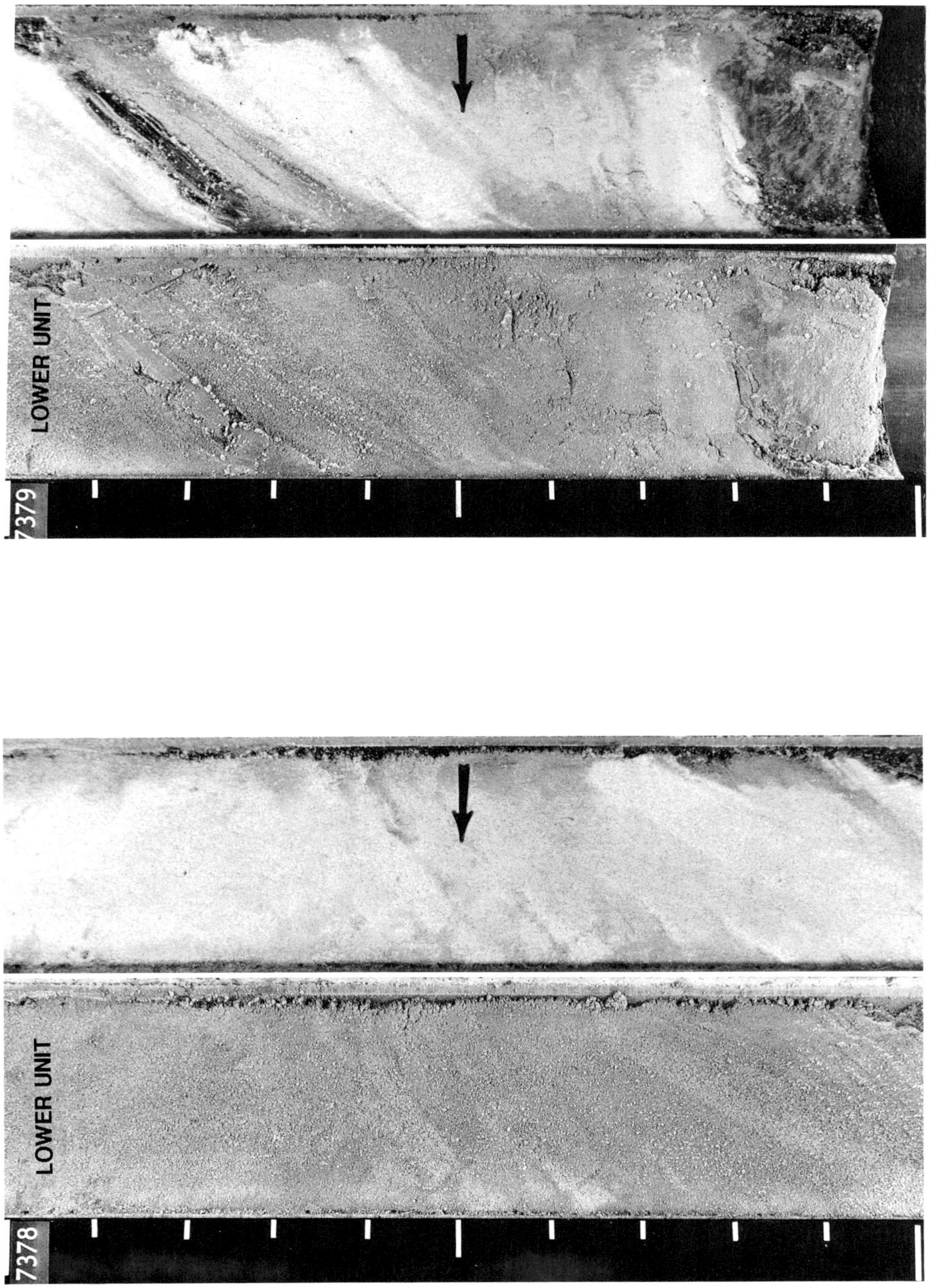

FIG. 11.—Whole core 7,378-7,380 ft MD showing laminated sand (white and grey on black light photo) with interbedded shale (black on black light photo) of lower flow unit of core 3 (see Fig. 13). Plane light on left, black light on right. Scale 5 1/2 in=1 ft. Footage marked in tenths.

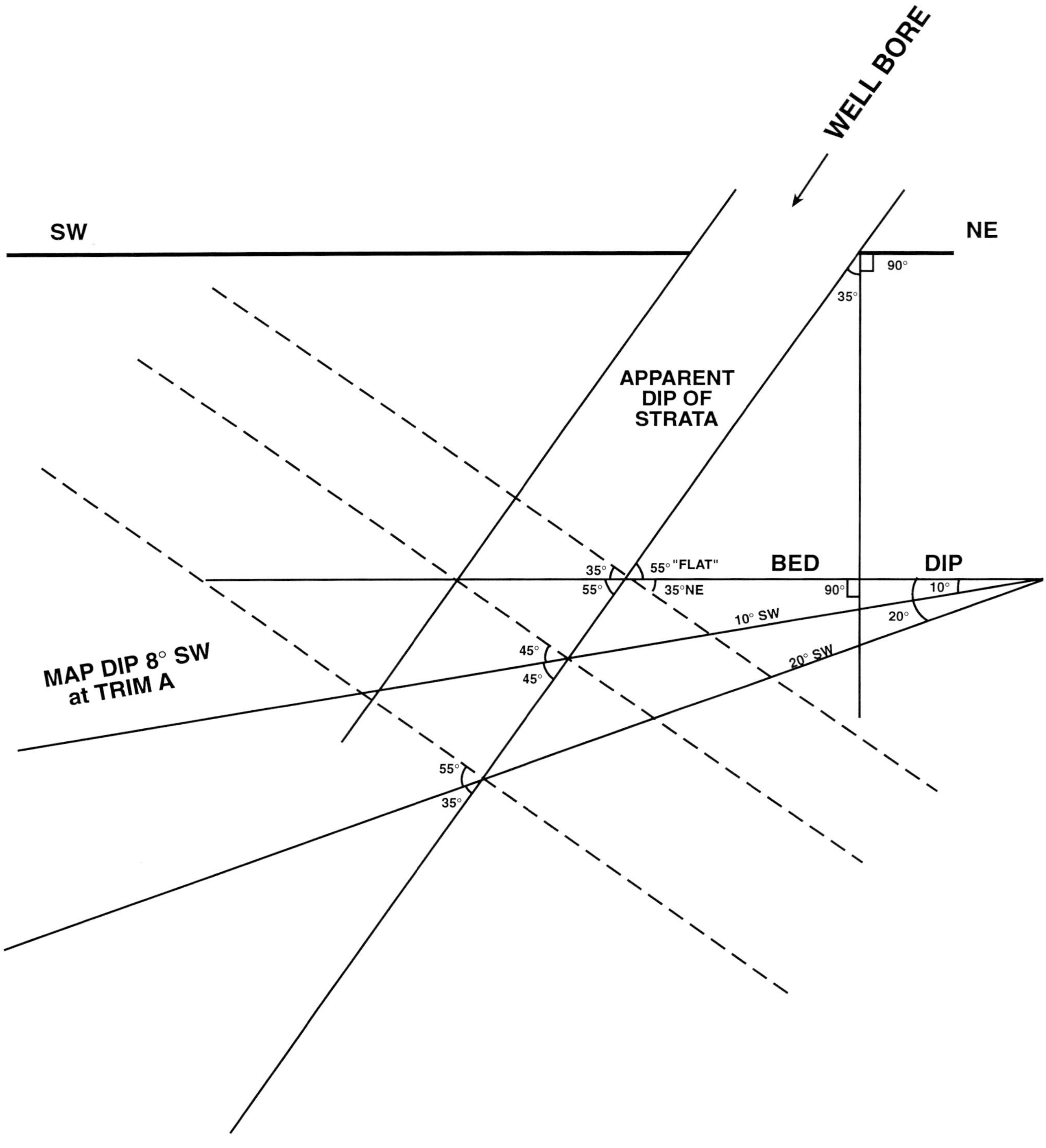

FIG. 12.—Diagram projected in the plane of GB 171 #2 borehole showing relationship of borehole deviation, bed dip and apparent dip of strata.

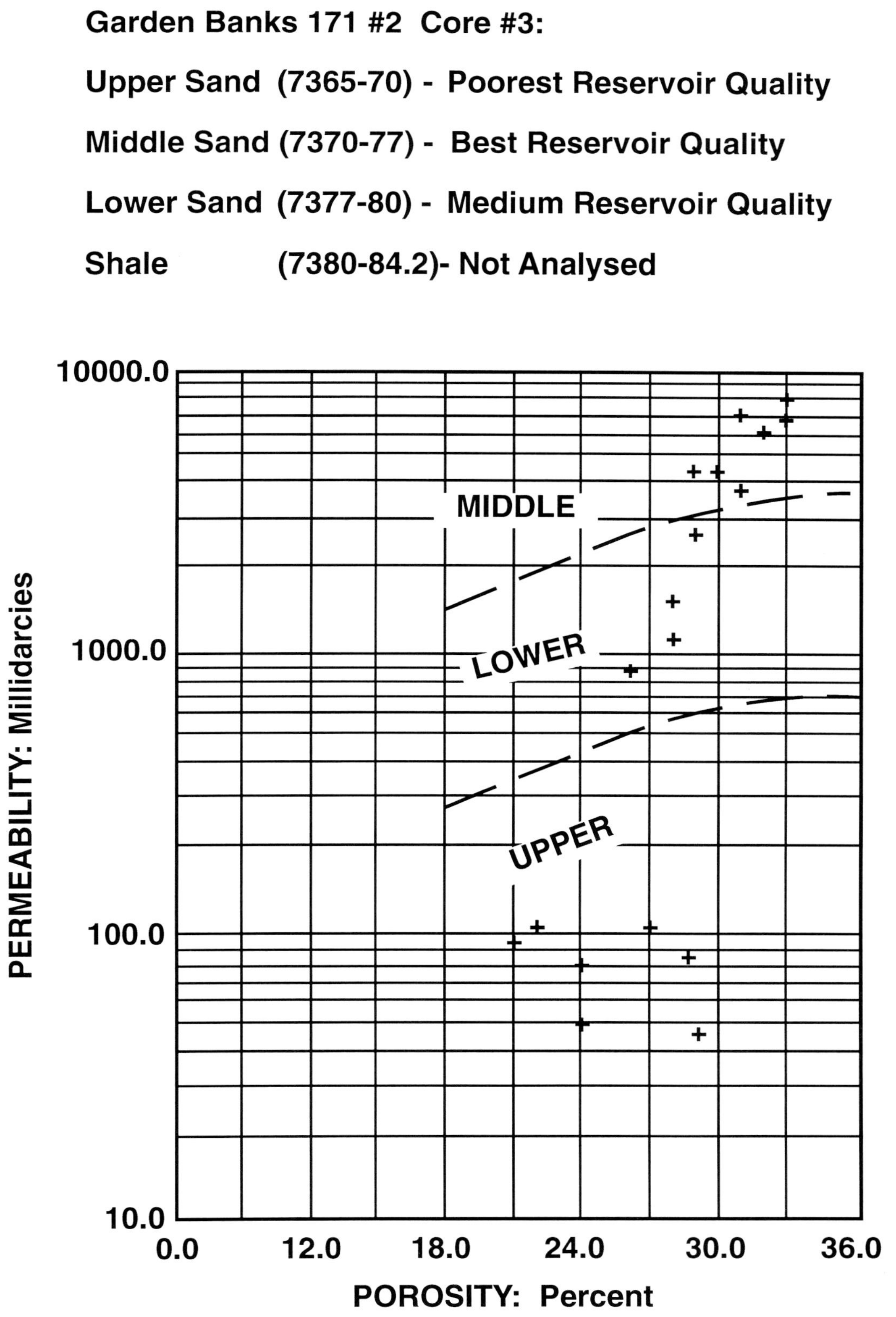

FIG. 13.—Plot of permeability versus porosity (measured under 2,000 psi confining pressure) of whole core plugs from core 3 showing the subdivision of reservoir quality into three flow units: lower, middle and upper.

the form of clean, very thin, continuous silt laminae alternating with clay-rich laminae. The silt laminae are made of angular to subrounded, very well sorted, silt-size grains with fair, intergranular porosity. Clay-rich laminae consist of: 54-61% mixed-layer illite/smectite, 20-22% illite, 10-16% kaolinite and 5-11% chlorite (percentages normalized to 100%).

Sedimentary strata in Cores 1 and 2 appear disturbed (Figs. 6 and 7). Apparent dips, 60 degrees to almost vertical, are 25-55 degrees after borehole deviation correction (Fig. 12) assuming a southwest direction. These structural attitudes are significantly more than can be accounted for by seismic (Figs. 1-3). The *Trimosina A* seismic horizon dips 8 degrees southwest. In addition, the finely bedded mud and silty sand in these cores show fine-scale folds, soft-sediment shearing, and non-horizontal, closely spaced fractures. These features could not be detected with sidewall cores. These structures and sediment dips are not interpreted as drilling or coring disturbances. Rather, these features are consistent with down-slope creep and other types of shear or slumping. Wireline logs do not indicate pay in this section, however, the cores show potential for fair reservoir quality, although continuity is not likely in these disturbed sediments. Once the FMS images were calibrated with the whole core, the FMS images above and below the core were investigated. Hundreds of feet of high dips and laminated sands and silts are present throughout the *Trimosina A* zone similar to Cores 1 and 2 in the well indicating the possibility of many feet of pay not indicated on the wireline logs.

Core 3 From Second Zone

Core 3 (Figs. 8-11) is from a sandier interval made up of unconsolidated, interlaminated silty sand and massive medium grained sand (up to 180 micrometers). The silty sand within the laminae is silt- to coarse sand-size, well sorted and subangular to well-rounded and has fair intergranular porosity. The massive sand has exaggerated pore geometries indicating considerable textural modification which is probably a result of decompression of the core. The sand is angular to well rounded, moderately well to well sorted, silt- to coarse sand-size with well-developed intergranular porosity. Its matrix contains 1-9% detrital clay with nearly the same composition as the clay-rich laminae of the laminated silts from Cores 1 and 2.

The sand in Core 3 can be divided into three flow units as shown on a plot of permeability versus porosity (Fig. 13). The three units are lower, middle and upper with the middle unit having the highest permeability and porosity and the upper unit having the lowest. The average porosity and geometric mean permeability measured under 2,000 psi confining pressure from sand in Core 3 is 27.% and 671 md, respectively.

The predominant apparent dip measured in Core 3 is 50 degrees or 15 degrees corrected for borehole deviation assuming a southwest direction (Fig. 12). Dips noted on Figure 5 are corrected for borehole deviation. There can be error in core dips as core rotation frequently occurs along coring induced fractures. Care was taken to discount dip measurements and dip reversals across these zones. Beds in Core 3 show horizontal, inclined, or contorted laminations; some sands are massive. Thicker sands have undergone soft-sediment deformation (Fig. 9). Evidence as to whether this interval is slumped or if contortions are due to coring is more equivocal than for Cores 1 and 2. The approximately 15 degree dip of the unit, however, suggests the zone is disturbed. Original depositional dip may be inferred to be similar to the modern upper slope surface which dips 1-2 degrees as evidenced by bathymetric maps of this area.

Primary sedimentation structures in all three cores are too deformed to suggest much about depositional environments; however, structures including evidence of slumping are consistent with a upper to mid-slope setting as suggested by paleobathymetry from benthic foraminifera. The slump structures and rotated blocks may also be related to the nearby faults apparently active during deposition of the middle Pleistocene.

DIP ANALYSIS

High apparent dips measured from the whole core inspired detailed dip analysis. Routine processing of dips from the FMS was not able to resolve the apparently adequate dip information acquired from this well. Detailed investigation of the FMS images resulted in parameters adequate to process the dip information such as higher search angle and a shorter correlation step. Structural attitudes were consequently resolved.

Structural attitudes calculated from the FMS dip analysis are widely scattered. The FMS images show that while computed dips are somewhat chaotic overall, there are intervals of tens of feet with consistent dips throughout the well, as the whole core samples, so other coherent, but rotated blocks of laminated fine-grained sands and silts may occur.

LOG ANALYSIS

Evaluation of wireline logs and sidewall cores analyses estimated 300 net ft of oil pay between 4,900-8,900 ft in the GB 171 #2 well. Processed FMS images were calibrated with the core and used to visually confirm pay. Evaluation of the FMS images suggest the possibility of more feet of pay in zones of low resistivity log values.

DRILL-STEM TESTS

The first two drill-stem tests (DST's) tested low resistivity pay with log characteristics similar to the intervals of Core 1 and 2. The first (DST) (8,330-8,380 ft MD) was attempted through 9 5/8" casing and 7 5/8" liner without a gravel pack. The well did not flow. A trace of oil and 3 barrels of calcium bromide and seawater with a few gas bubbles was reversed out of the drill pipe after the test. Testing through two strings of pipe without a gravel pack left some doubt as to the reliability of the test. The second DST (7,530-7,605 ft MD) was interrupted by Hurricane Jerry. The well flowed 4 barrels per hour of 95% oil before being shut-in for the hurricane. After the storm blew over, the well was opened and flowed 512 MCFGPD and 3-4 barrels per hour with a trace of oil. The test produced 20 barrels of 31.1 degree API oil. This test's results are questionable because of the interruption due to weather.

The third and final DST, over the Core 3 interval, was a technically satisfying but economically disappointing test. A gravel-packed flow test of the interval 7,320-7,383 ft MD had 52 ft of measured pay and a true stratigraphic thickness of 35 ft. Whole core analysis indicated a clean sand, with porosity and permeability estimated at 27.9% and 671 md, respectively. This interval was the thickest sand with the best resistivity and porosity of any sand in the well.

The final DST was made up of three buildup and three flow periods. Analysis of the buildup data indicated that the reservoir experienced extreme pressure depletion during the test. The difference between the initial and final extrapolated reservoir pressure was approximately 2,500 psi. The maximum flow rate during this DST was 1548 BOPD. The final rate before the final shut-in period was 242 BOPD, 0 BWPD and 1654 MCFGPD with 240 flowing tubing pressure on an open choke.

The interpretation of this test is that the reservoir is very small, perhaps just a few thousand square feet, and could possibly produce at a sustained, but low rate. Low permeability, 35 millidarcies, was calculated from this test. Based on the core descriptions and FMS analysis the sand in Core 3 is probably a broken, rotated block of laminated and massive sand not well connected to other similar sand. This third DST provided evidence of reservoir discontinuity in the *Trimosina A* zone corroborated by the whole core. Without the whole core, an explanation of an extremely small, discontinuous sand body with low permeability would have been more conjectural.

CONCLUSIONS

Whole core from the Marathon #2 well in Garden Banks Block 171, Offshore Louisiana, Gulf of Mexico, was essential to integrating all of the well data of the *Trimosina A* mid to upper bathyal sandstones, siltstones and shales. Before the whole core from the GB 171 #2 was acquired, good rates were expected from the oil-filled sandy intervals, although continuity was a concern considering most of the pay zone was interpreted to be in a cut and fill sequence with poor seismic continuity. The whole core was a key to the conclusion that connectivity is a problem regardless of possibly questionable results from the drill-stem tests. The whole core was crucial to understanding poor reservoir potential from good reservoir quality sand and economically disappointing hydrocarbon recoveries from drill-stem tests. The whole core was also able to improve the dip analysis.

ACKNOWLEDGMENTS

I wish to thank Marathon Oil Company for allowing this material to be presented and for their support during manuscript preparation. I wish also to thank Robert D. Winn, Jr., (formerly of Marathon Oil Company) for his assistance and expertise in description of GB 171 #2 core. The manuscript has been greatly improved thanks to reviewers and co-editors Robert D. Winn, Jr. and John M. Armentrout and to helpful comments from Patricia A. Santogrossi.

REFERENCE

STEINMETZ, J. C., 1986, Cenozoic timescale and microfossil zonation: Marathon Oil Company, Bulletin B86-10R, 1 chart, 1 p.

PHYSICAL CHARACTERISTICS OF DEEP WATER THIN-BEDDED TURBIDITES, GREEN CANYON BLOCK 154, GULF OF MEXICO

NANCY TRUMBLY BENTHIEN
Marathon Oil Company, 5555 San Felipe Road, Houston, TX 77056

ABSTRACT

Twenty-nine feet (8.8 meters) of lower Pleistocene core material were recovered by Marathon Oil in its Green Canyon Block 154 #1 well, offshore Louisiana, in order to more fully evaluate the exploration program being drilled by Marathon and its partners in this deep-water locality of the Gulf of Mexico. From the mid- to late- 1980s, the partnership drilled seven wells within four miles of Shell Offshore Incorporated's newly-discovered Bullwinkle Field. Oil and gas pay zones were penetrated in the Plio-Pleistocene of these wells, and Marathon took the core to complete a reservoir characterization study that included evaluating wireline logs, drill stem tests and extensive seismic modelling in addition to the core analysis.

The cored sediments are part of a larger unit of thinly bedded, fine-grained turbidites deposited in lower bathyal water depths on the edge of a Plio-Pleistocene salt withdrawal sub-basin. This interval is comprised of base-absent Bouma sequences (stacked T_b-T_e cycles), and contain parallel planar, undulose and rippled laminations. Sand beds are no more than five inches thick. The cored interval has a characteristic log signature for low contrast/ low resistivity pay commonly found in laminated silt/sand/shale lithofacies. Drill stem tests of this unit yielded an average flow of 1391 barrels of oil/day and 1040 thousand cubic feet of gas/day on a 20/64 inch choke.

INTRODUCTION

Lower Pleistocene deep-water sediments were cored by Marathon Oil Company in its Green Canyon (GC) Block 154 #1 well on the present-day slope of the Gulf of Mexico (Figure 1). The well is located approximately 220 miles (354 kilometers) south- southwest of New Orleans, Louisiana in 1410 feet (430 meters) of water. Drilled in 1988, the GC Blk. 154 #1 well was part of an extensive exploratory drilling program by Marathon and its partners, Amerada Hess, Transco Exploration, and Sun Exploration. Six other exploration wells were drilled by the partnership in the immediate vicinity, but none of the other wells were cored.

The conventional core was cut from the interval 7472-7518 feet (2277.5-2219.5 meters) with recovery from 7472-89 feet (17 feet or 5.2 meters) and 7497-7509 feet (12 feet or 3.6 meters). Log analysis calculated 49 feet (14.9 meters) net hydrocarbon pay in the well (Figure 2). After casing the borehole, an interval was perforated from 7458-7510 feet (2273.2-2289 meters), gravel packed, and production tested in order to evaluate productivity of the thinly bedded and laminated low resistivity sand encountered in the well.

This paper focuses on the sedimentological, petrological, and petrophysical character of the cored interval and on how these parameters relate to depositional environment and hydrocarbon producibility.

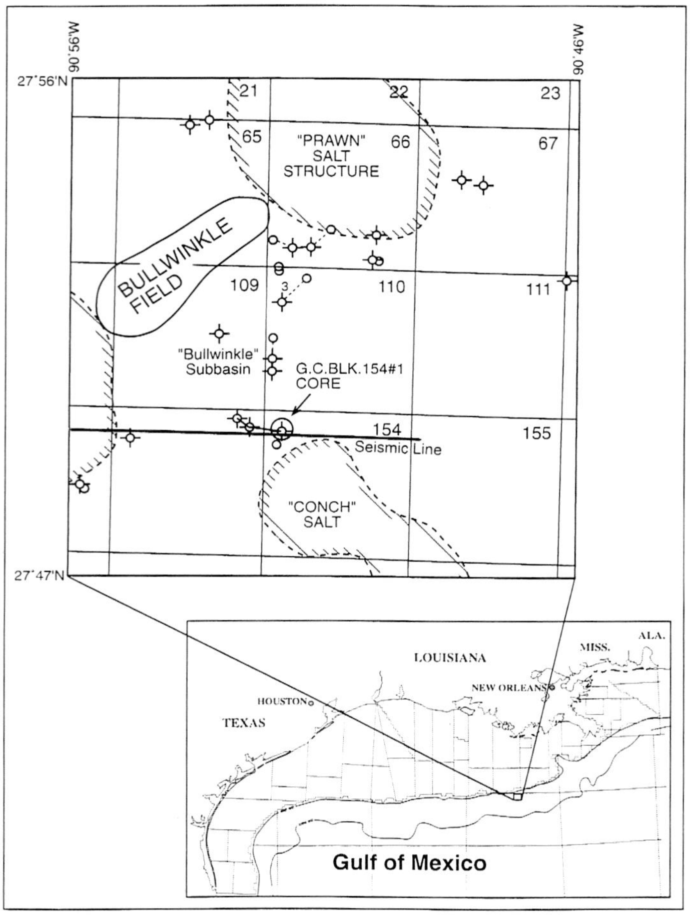

Fig. 1.—Location of core study.

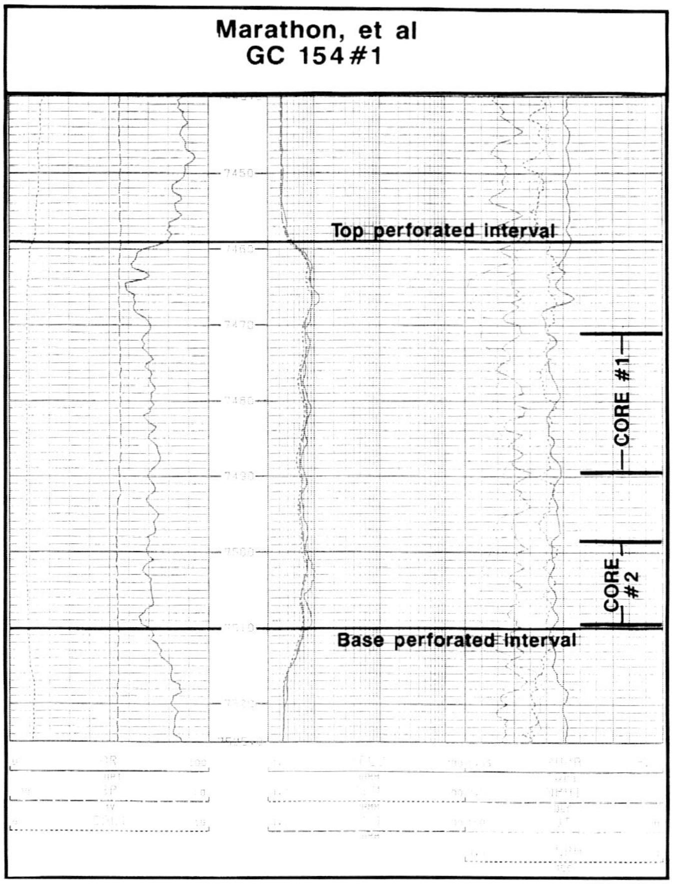

Fig. 2.—Log over the lower Pleistocene pay zone. The cored and perforated intervals are shown.

GEOLOGIC SETTING

The GC 154 #1 well was drilled on the southeastern edge of the "Bullwinkle" salt withdrawal sub-basin. Bullwinkle Field is located approximately 3 miles away on the northwestern edge of the same sub-basin but produces from an older, early Nebraskan submarine fan system (Holman and Robertson, 1994). Plio-Pleistocene sedimentation in the "Bullwinkle" sub-basin caused uneven loading and dissection by resedimented clastics of a buried allothonous salt canopy (West, 1989; Diegel, et al, 1993) (Figure 3). Remnants of the salt canopy formed topographic highs on the paleoslope between which thick sequences of sediments were transported and deposited. Continued loading of the mobile substrate caused further subsidence and deposition. The west-to-east seismic line in Figure 4 shows the thickening of Plio-Pleistocene sediment along the axis of the "Bullwinkle" sub-basin and thinning toward the "Conch" salt feature. The interval topped by the 1.1 Ma sequence boundary was cored in the G. C. 154 #1 well. Seismic character within this sequence shows nested reflectors in the basin axis. These reflectors overlap the basin margin, and represent the focal point of turbidity deposition.

Paleontological analysis places the cored interval in lower bathyal water depths between the extinction points of the calcareous nannofossils *Discoaster brouweri* and *Calcidiscus macintyrei* (1.90-1.45 ma b.p.) in the early Pleistocene (Steinmetz, 1988). This age determination is consistent with regional mapping of the seismic stratigraphic sequences in the area.

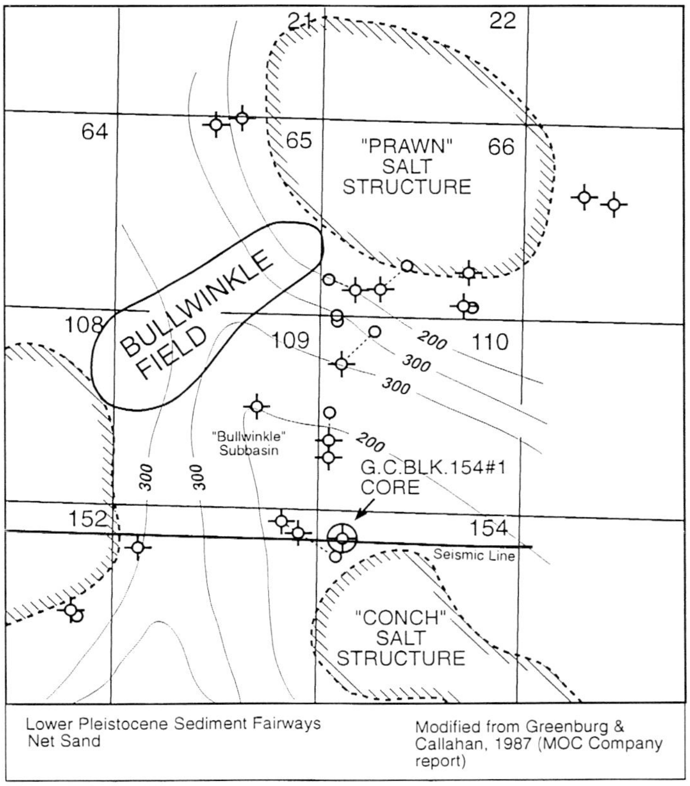

Fig. 3.—Sediment fairways in the Bullwinkle salt withdrawal sub-basin. Sediment was transported and deposited in the low-lying areas between salt features as indicated by sand "thicks".

CORE DESCRIPTION

The sandy unit from which the core was taken exhibits a moderately serrate to blocky gamma ray log motif with a gradational base indicating progradation of this turbidity system over basinal muds (Figure 2). The top of the unit gradually fines upward into mud thus completing turbidity deposition. Whole core taken in the middle of the unit consists of base-absent Bouma sequences (stacked T_b-T_e cycles) and is interpreted as a series of thinly bedded, fine grained turbidites (Bouma, 1962). The core is poorly consolidated and highly laminated as shown in the ultraviolet light prints in Figure 5a and 5b. The sediment is composed of mud, silt, and very fine grained sands which are interlaminated to thinly bedded in a regular, repetitive manner. The laminae are parallel planar, undulose and rippled (Figure 6). Contacts between laminae tend to be sharp (e.g. 7,504 ft., Figure 6). Occasional load structures are present at the interface between mud and some of the thicker sand beds (Figure 7).

The sand beds tend to be massive, occasionally graded, and no thicker than 5 inches (7,500.4 ft., Figure 7). Grading within individual sand lamina is observed (Figure 8). Contorted bedding indicative of soft sediment deformation is present at several intervals in the core such as at 7,498.5 ft. (Figure 9a).

Bioturbation appears to be minor, but is demonstrated at 7478.8 ft. by fine-scale disruption of laminae (Figure 9b). Other biological activity is noted by the presence of fecal pellets in some of the mud laminae, as well as sand-filled burrows at sand/mud lamina contacts (Figure 10). The preservation of such delicate structures indicate a relatively stable environment punctuated by occasional low energy turbiditic flows depositing the base-absent Bouma sequences that are observed.

PETROGRAPHY

The analyzed samples consist of poorly consolidated mud interlaminated with unconsolidated silt and very fine grained sand. The shale laminae may be massive or pelleted (Figures 10 and 11 respectively). Pelleted mud decreases in abundance with increasing depth suggesting longer periods of quiescence in the younger sediment.

The sandy silt and silty sand laminae are feldspathic litharenites. Framework grains consist primarily of quartz (44-64%), sedimentary rock fragments (14-27%), and feldspar (8-14%). Clasts are generally subrounded and equant in shape (Figure 12). Primary intergranular porosity is observed in these unconsolidated sediments and pore systems are believed to be well connected (Figure 13).

INTERPRETATION

All of the characteristics described above are consistent with an interpretation of thin-bedded turbidites, specifically, Bouma sequences composed of the B, C, D, and E divisions. The uniformity of the log above, below, and within the cored interval allows for the same interpretation to be given for the entire unit. The ubiquitous presence of base-absent Bouma sequences along with rare slumps and minor scours is evidence of episodic sedimentation along a layered sheet margin (Mahaffie, 1994). These turbidites may be interpreted as more distal equivalents to full Bouma sequences (Stow and Piper, 1984). However, further regional work remains to be completed to locate these

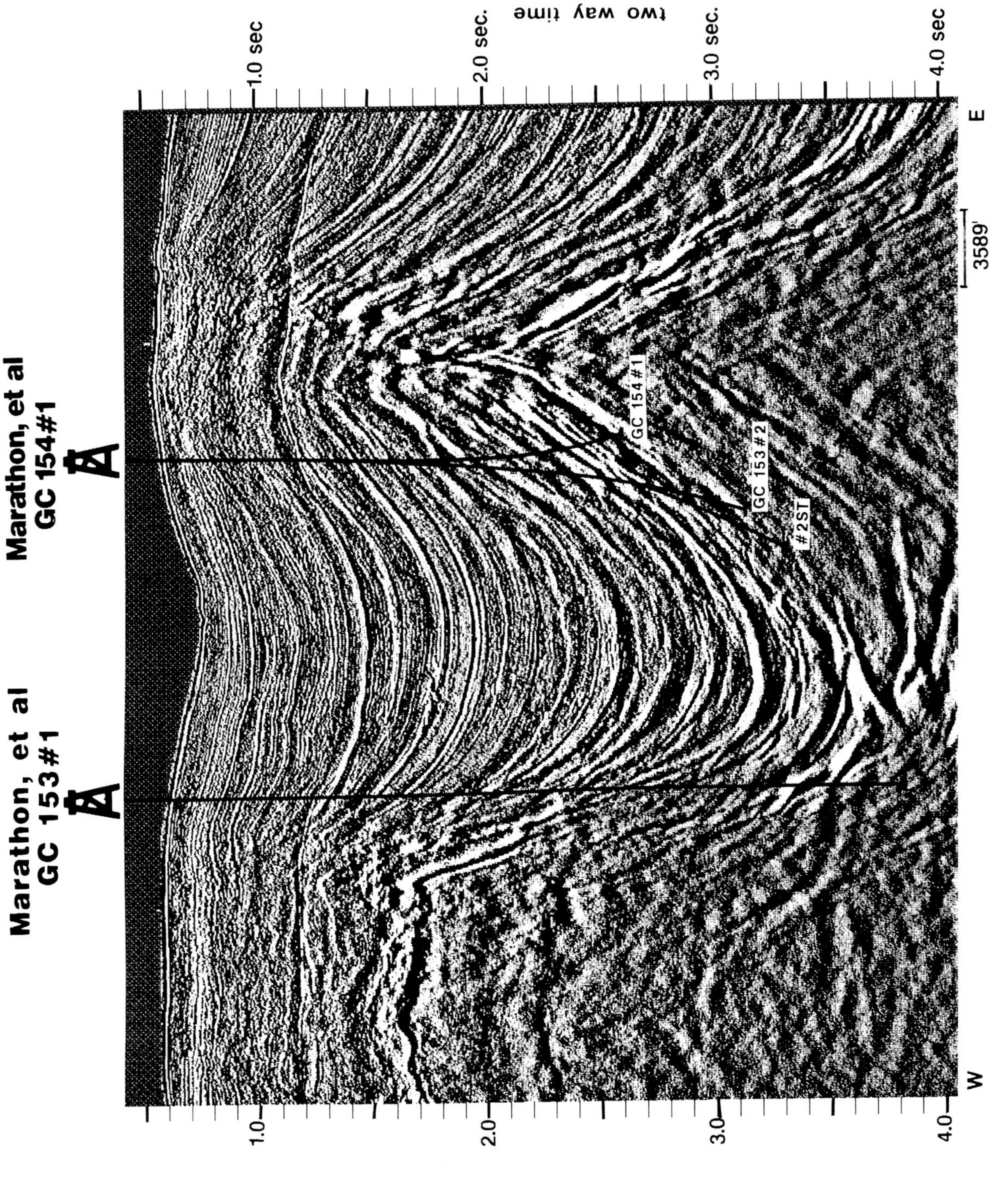

Fig. 4a.—Uninterpreted section. West to east variable density seismic display showing an axis of Plio-Pleistocene deposition between salt cored topographic highs.

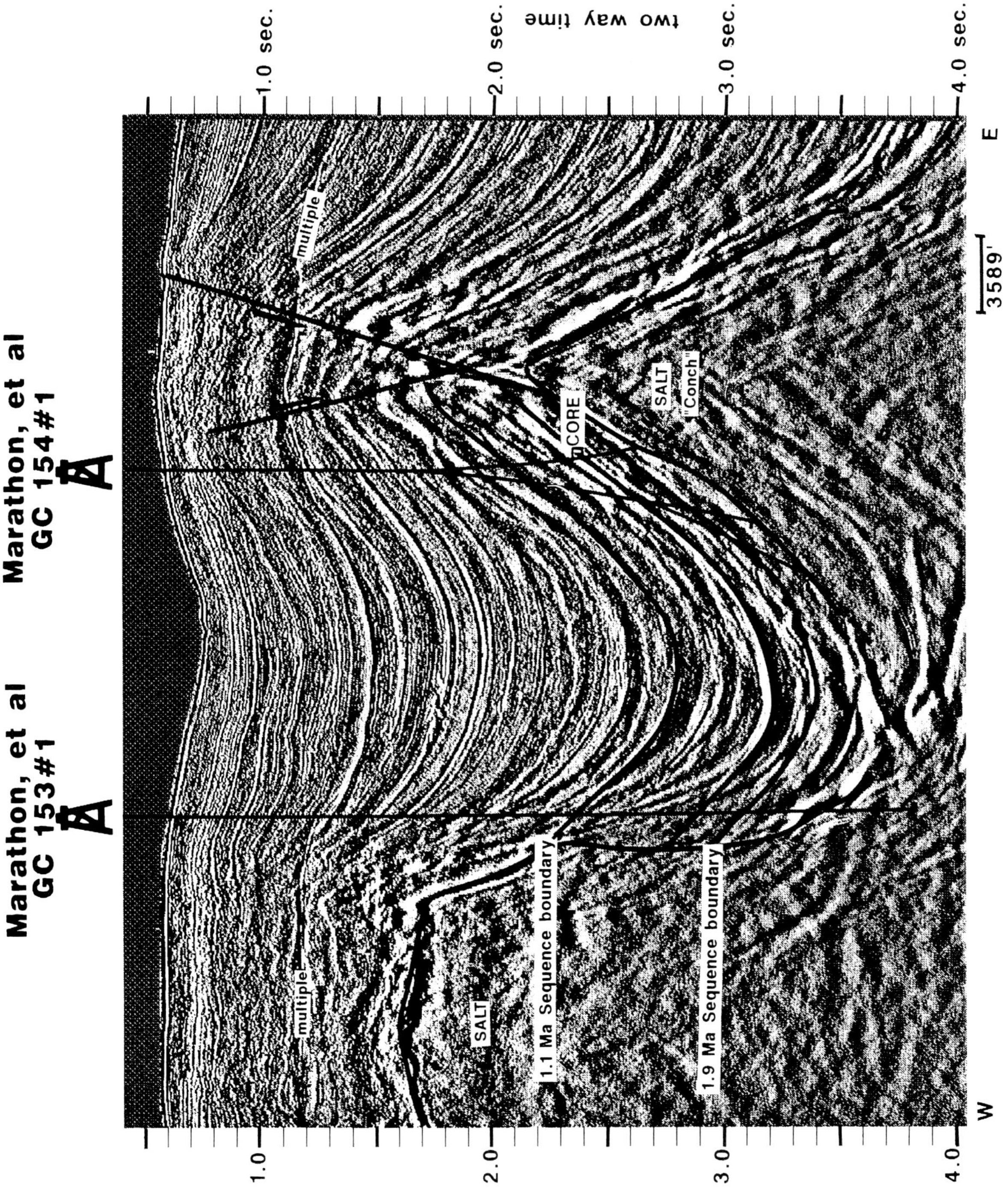

Fig. 4b.—Interpreted section. West to east variable density seismic display showing an axis of Plio-Pleistocene deposition between salt cored topographic highs.

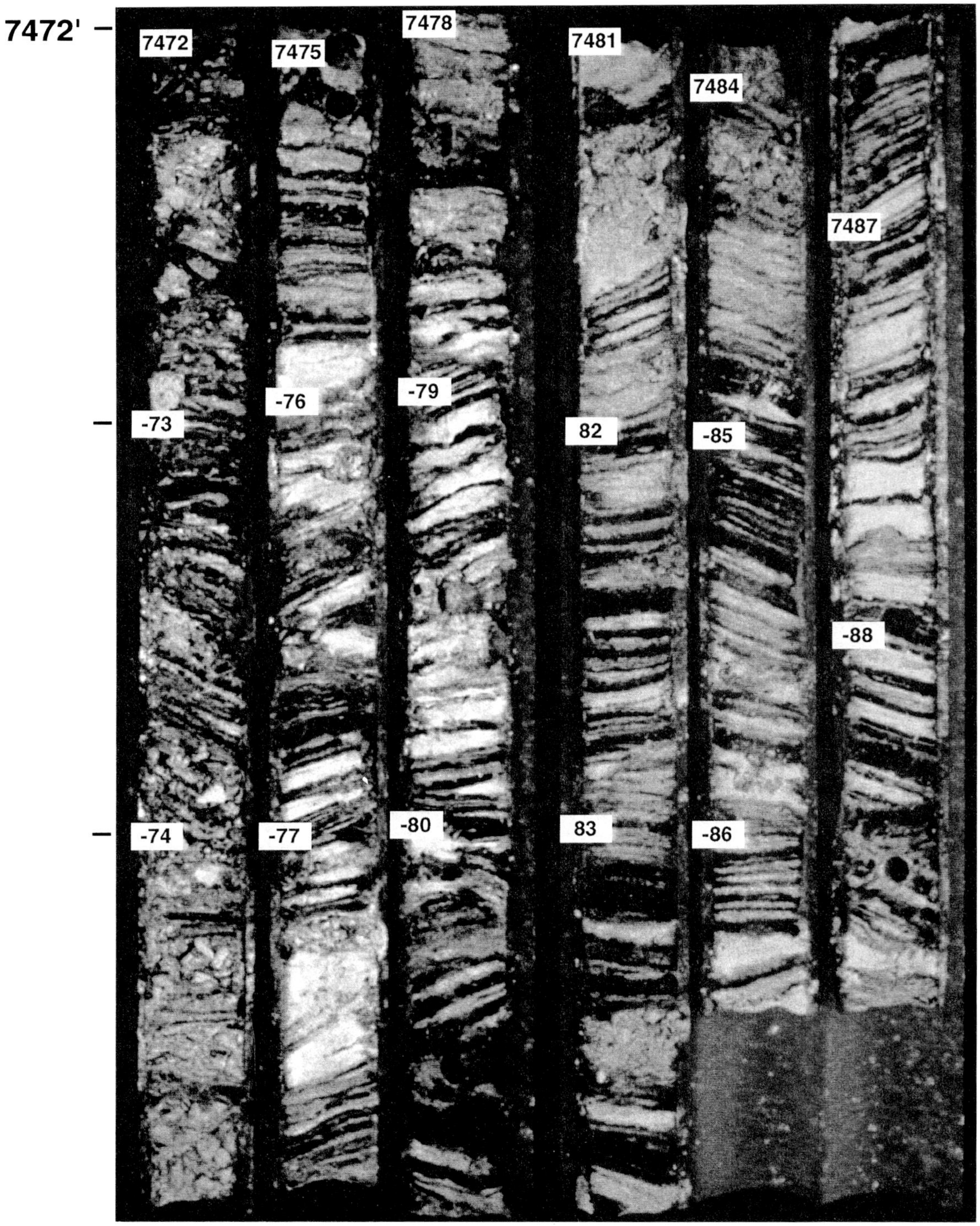

Fig. 5a.—Recovered core between 7,472-7,488 ft. Core slab shown in ultraviolet light. Sand laminae are light, shales are dark. Due to the lack of contrast between shale and oil-stained sands, white light prints are not shown.

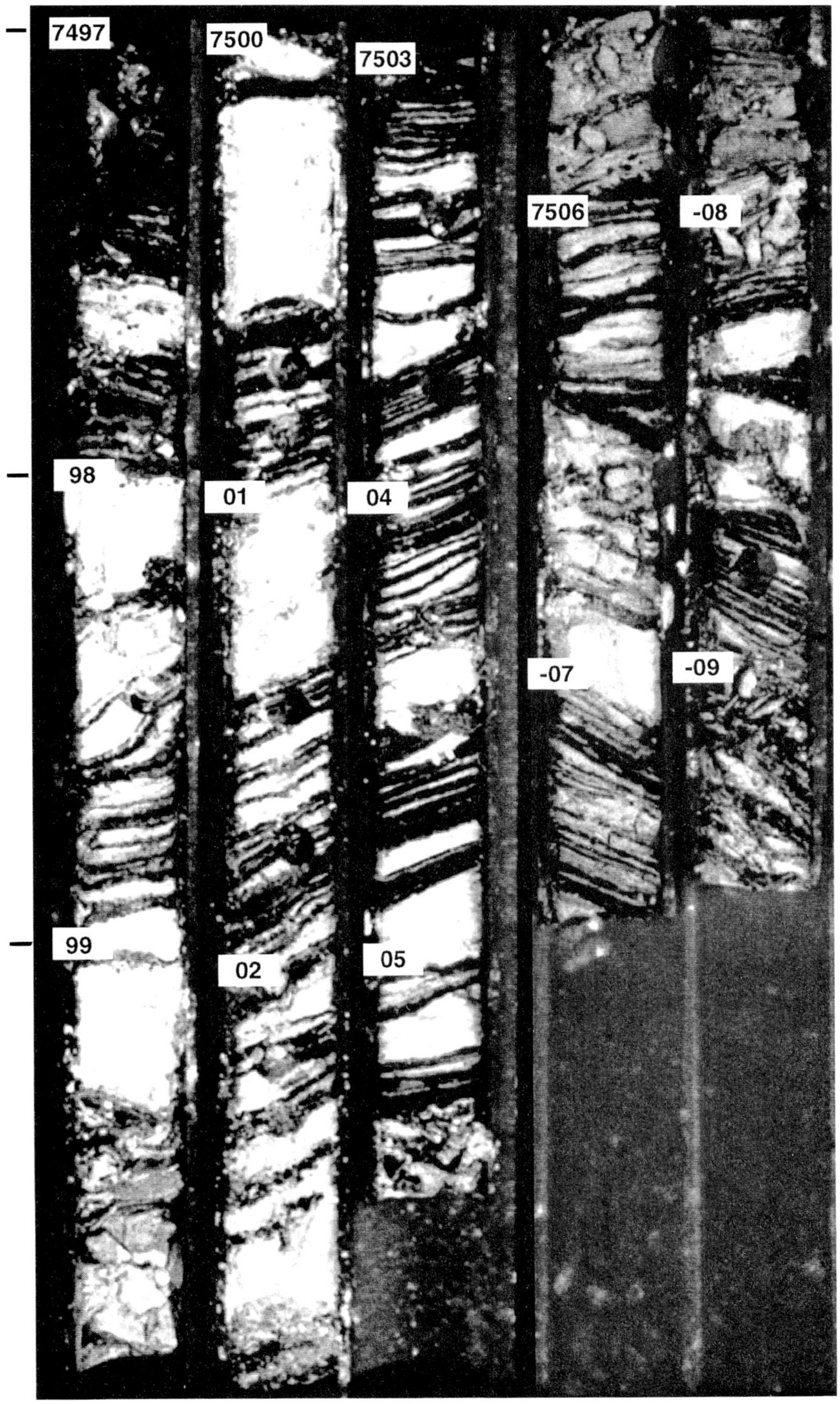

Fig. 5b.—Recovered core between 7,497-7,509 ft. Core slab shown in ultraviolet light. Sand laminae are light, shales are dark. Due to the lack of contrast between shale and oil-stained sands, white light prints are not shown.

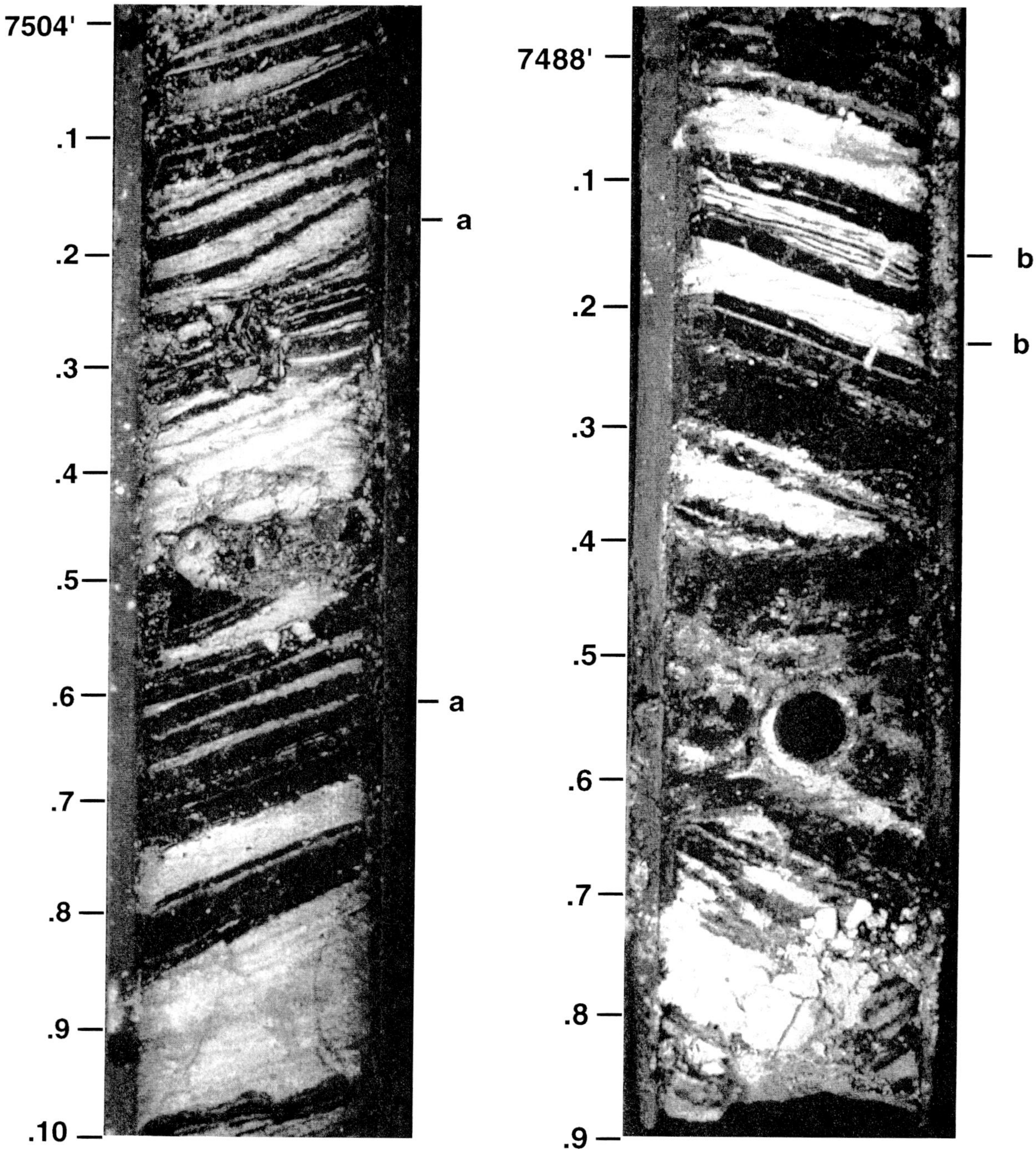

Fig. 6.—Core photographs in UV light from 7,504' and 7,488'. a: undulose and planar laminae; b: rippled laminae. Note typical sharp contacts between sand and shale laminae.

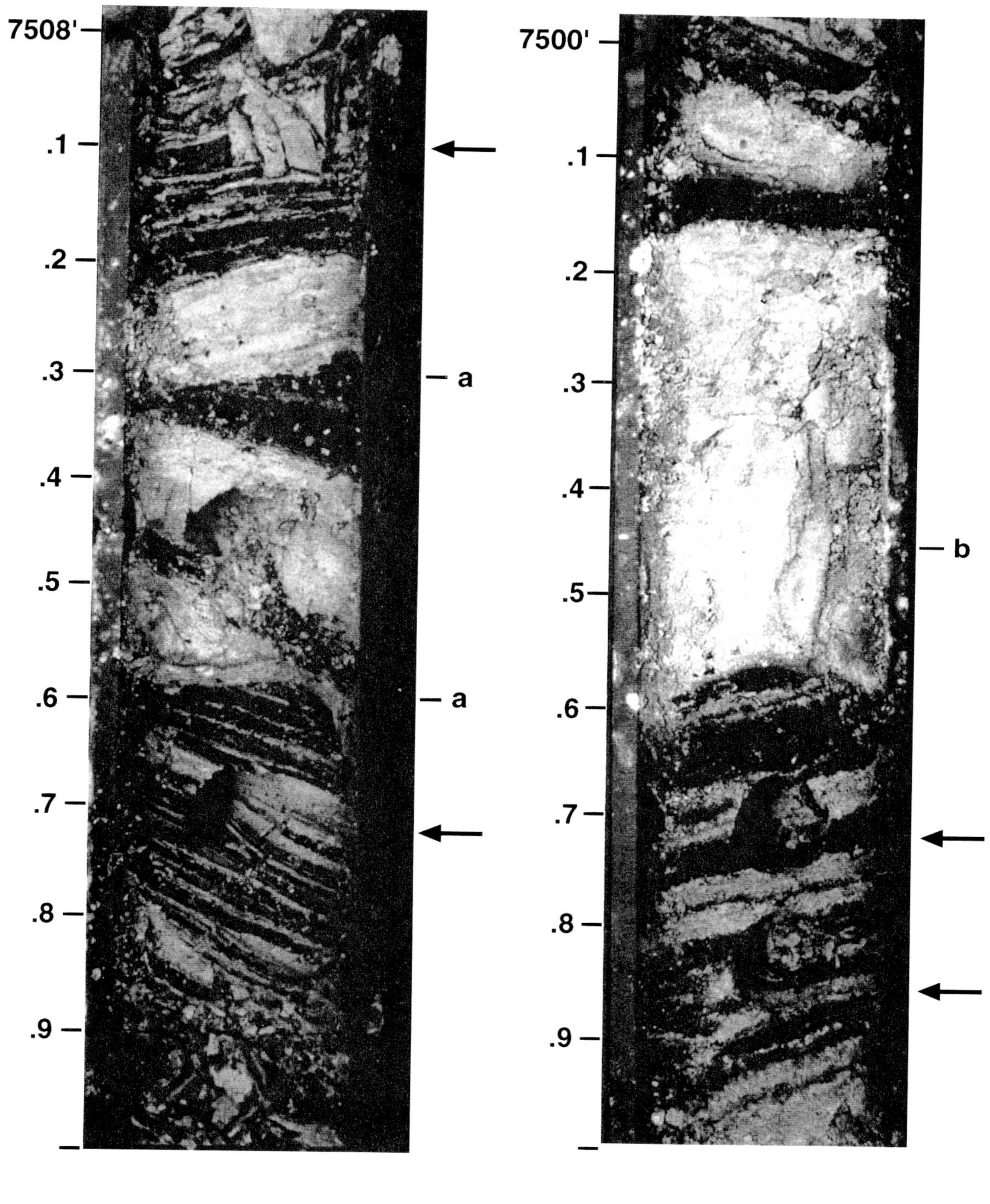

Fig. 7.—Core photograph in UV light from 7,508' and 7,500' showing load structures (a) at the base of sand beds and the thickest sand bed (b) in the core. Note lack of sedimentary structures in the sand. Core plugs at arrows.

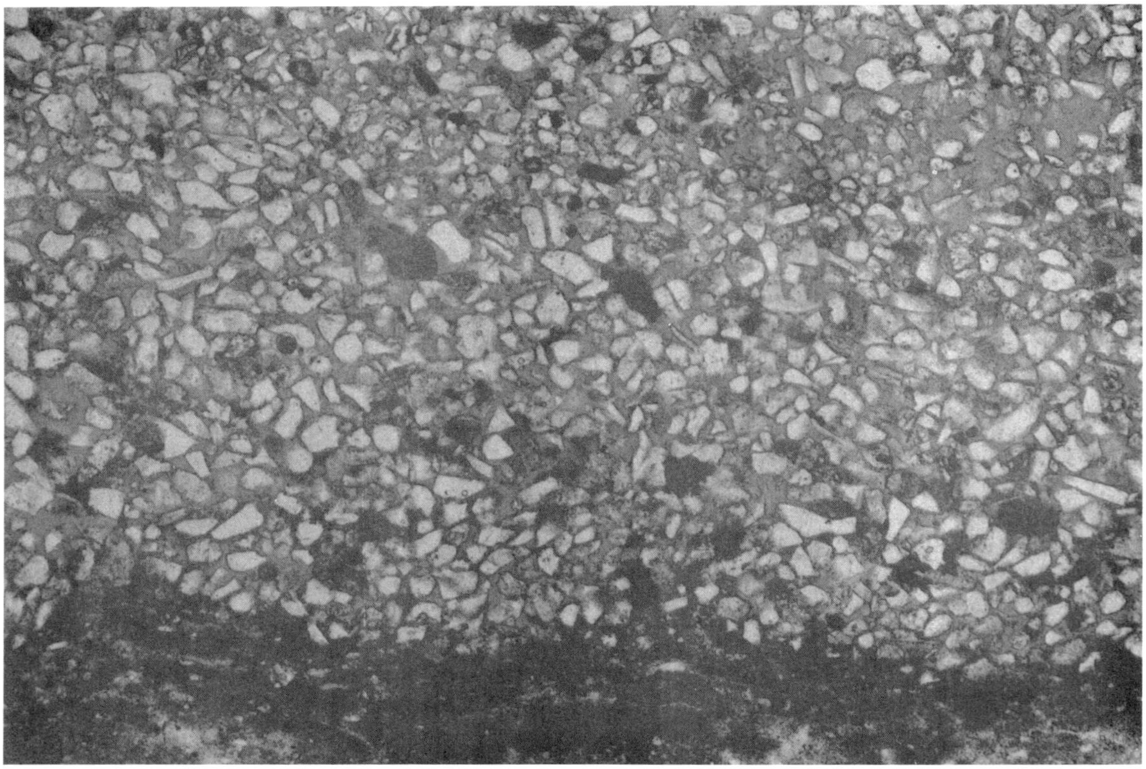

Fig. 8.—Photomicrograph from 7,506.9'. Note the sharp contact between sand and mud laminae and the grading within the sand unit (52x, plane polarized light).

fine-grained turbidites within the fan system. Thicker bedded turbidites may have been deposited to the north and west in the sediment fairway areas between salt structures (Figure 3).

PETROPHYSICAL CHARACTERIZATION

Figure 2 displays a basic log suite over the zone of interest, including the SP, gamma ray, resistivity, density and neutron porosity, and bulk density curves. In a quick-look interpretation, the log responses are not impressive. The gamma ray indicates that fairly abundant radioactive minerals are present. Resistivity measurements are less than 1.5 ohm-meters throughout the unit which is a typical response to hydrocarbons in a laminated silt/sand/shale lithofacies (Shelton and Hill, 1993). The density and compensated neutron porosity curves are widely separated, suggesting the presence of shale. The resolution of both of these tools is above the laminae thickness, thus they are unable to accurately measure porosity of sand in the interval.

The epithermal neutron porosity curve more closely follows the density porosity measurements suggesting reservoir porosities in the 30-35% range which is somewhat higher than the 21-31% porosity measured from the core. After careful log and core analysis, 49 feet (gross) of the 51 foot unit was calculated to be producible (Allen, 1988).

Drill stem tests supported the producibility of this laminated zone. During the DST the perforated interval attained an average flow rate of 1391 BOPD and 1040 MCFGPD on a 20/64 inch choke.

CONCLUSIONS

1. Fine grain deep water sediments were cored by Marathon Oil Company in its Green Canyon Block 154 #1 well in the Gulf of Mexico. These sediments

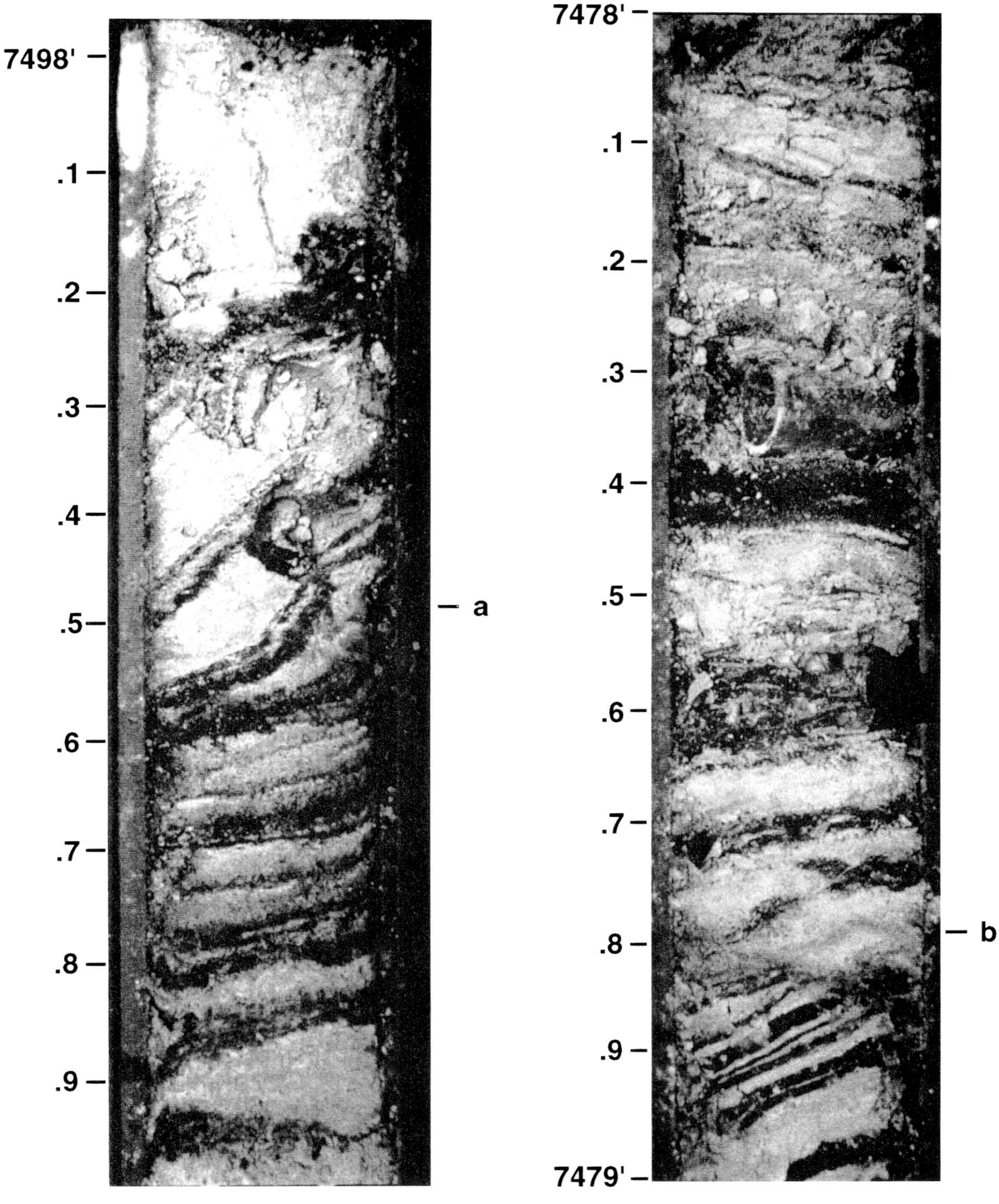

Fig. 9.—Core photograph in UV light from 7,498' and 7,478' showing soft sediment deformation (a), and showing possible bioturbation (b).

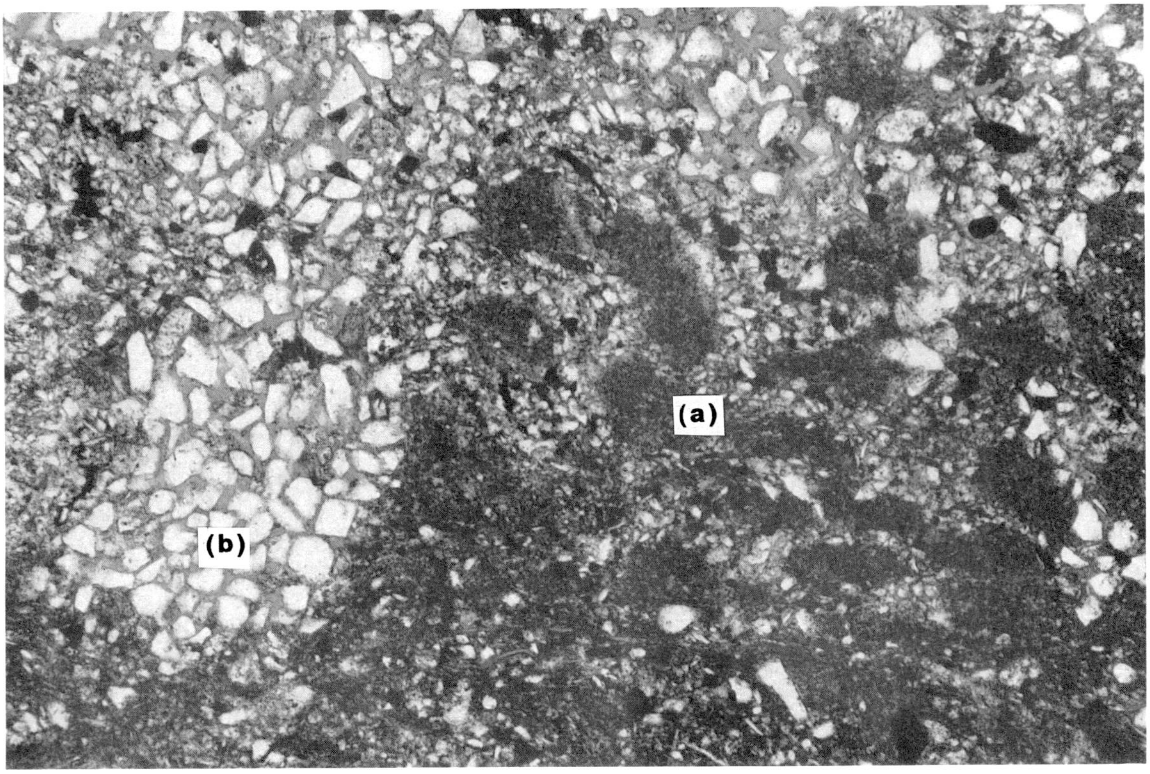

Fig. 10.—Photomicrograph from 7,488.8' showing pelleted mud (a) and sand-filled burrow (b) (52x, plane polarized light).

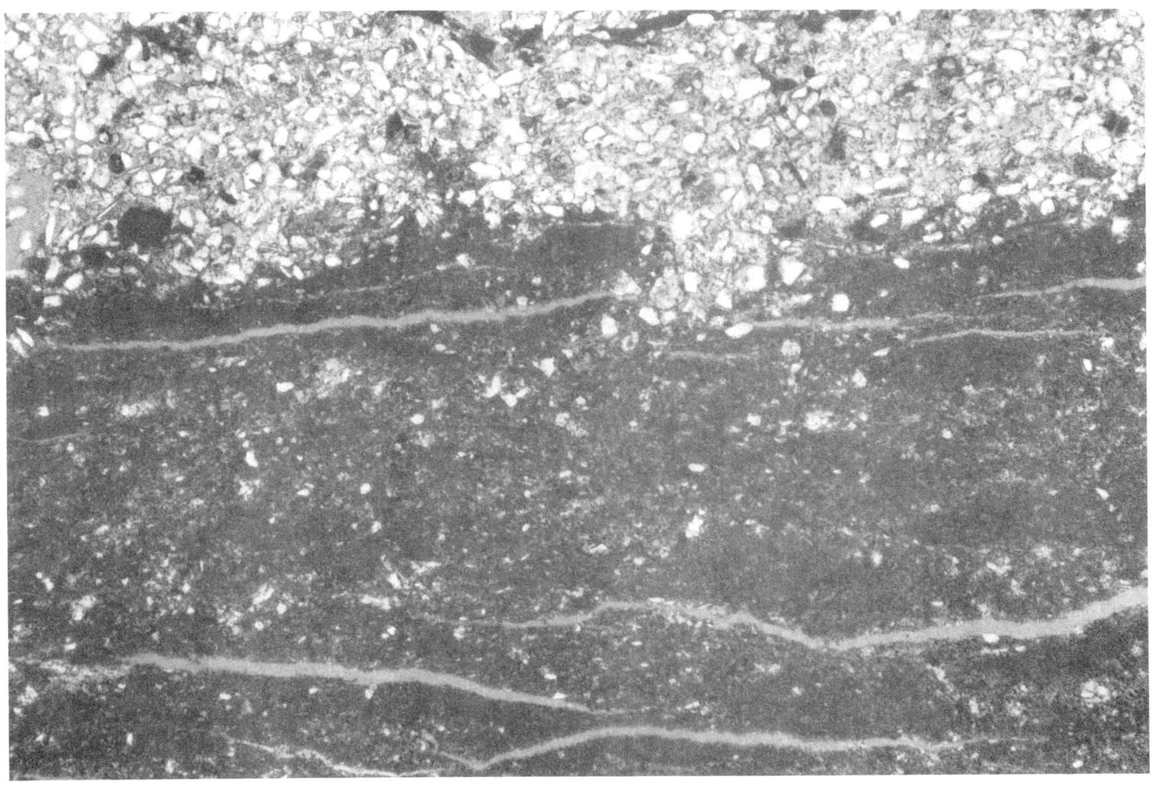

Fig. 11.—Photomicrograph from 7,508.8' showing structureless shale lamina in contact with sand bed. Fractures are coring induced (52x, plane polarized light).

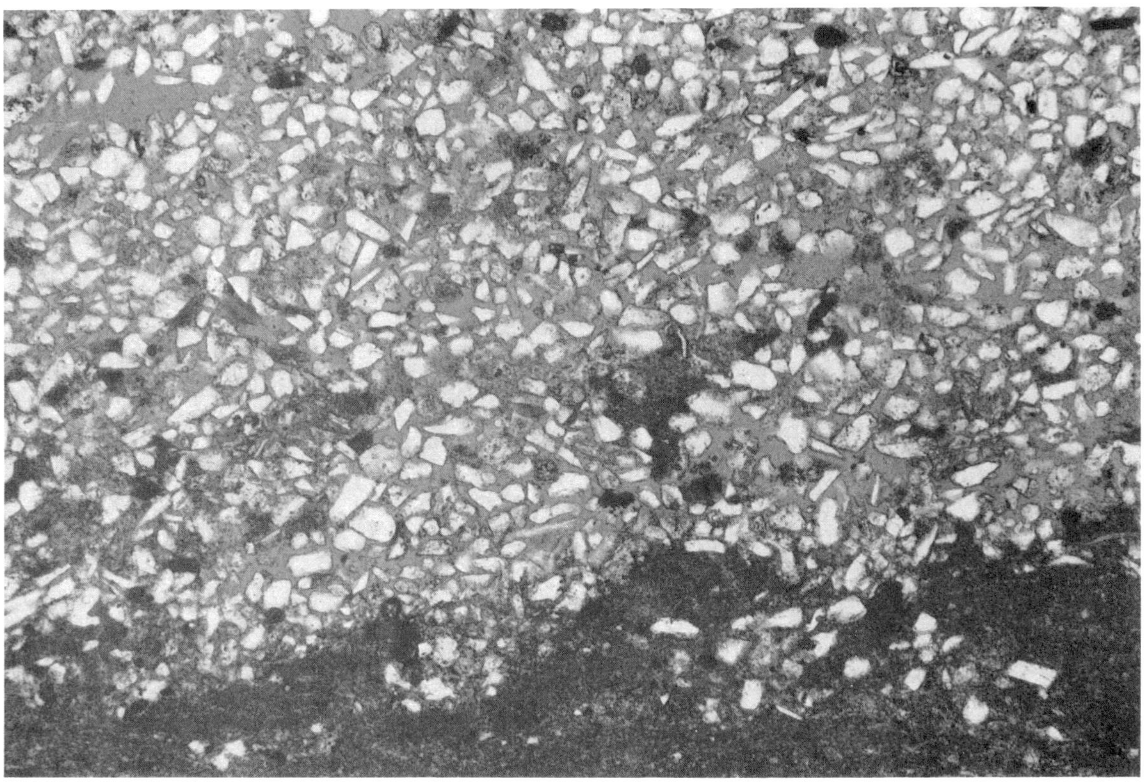

Fig. 12.—Photomicrograph from 7,503.8'. Very-fine grained sands and silts are generally subrounded and equant in shape (52x, plane polarized light).

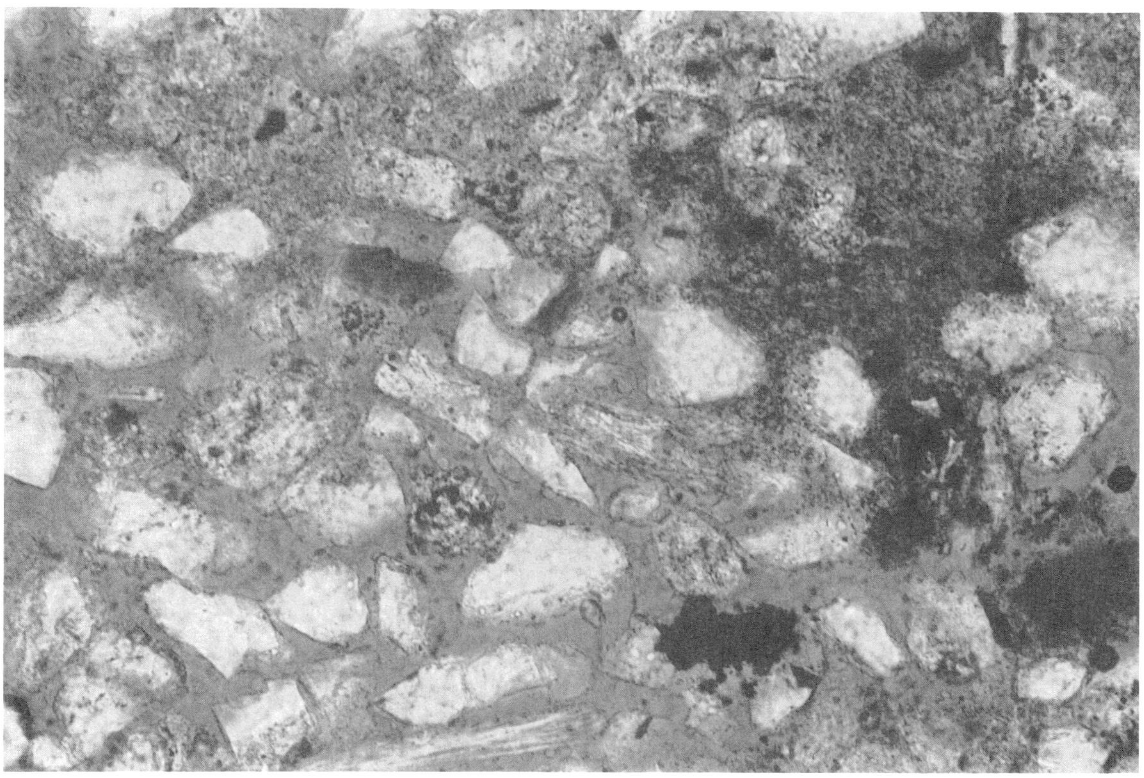

Fig. 13.—Photomicrograph from 7,503.8', enlarged scale. Grains are unconsolidated and retain primary intergranular porosity. Contacts are generally tangential (208x, plane polarized light).

were deposited in lower bathyal water depths on the edge of a Plio/Pleistocene salt withdrawal sub-basin created by the loading of a salt canopy.

2. The core contains parallel planar, undulose, and rippled laminations and thin beds composed of mud, salt, and fine-grained sands. Individual sand beds are no more than five inches thick.

3. These sediments are interpreted as fine-grained, thin-bedded turbidites comprising repetitive, base-absent (T_b-T_e) Bouma sequences. It is suggested that these turbidites represent deposition in a layered sheet margin with the focus of deposition being to the north and west along the axis of the sub-basin.

4. The cored interval has a characteristic log signature for low contrast, low resistivity pays, commonly found in laminated silt/sand/shale lithofacies. Drill stem tests of this unit yielded an average flow of 1391 barrels of oil per day and 1040 thousand cubic feet of gas per day on a 20/64 inch choke.

ACKNOWLEDGEMENTS

Thanks to John Armentrout, Ross Benthien, Patricia Santogrossi, and Robert Winn for helpful critical reviews of this work, and to Mike Anderson and Lara Malone for technical support in preparing the manuscript. I am thankful to Marathon Oil Company for granting permission to publish this paper.

REFERENCES

ALLEN, J. 1988. Marathon Oil Company internal report.

BOUMA, A. H. 1962. Sedimentology of Some Flysch Deposits. Elsevier, Amsterdam. 168 pp.

DIEGEL, F. A., KARLO, J. F., SCHUSTER, D. C., SHOUP, R. C., & TRAUVERS, P. R. 1993. Cenozoic structural evolution and tectonostratigraphic framework of the northern Gulf Coast continental margin, A.A.P.G.-S.E.P.M. 78th Annual Convention Abstracts, p. 91.

GREENBURG, C., & CALLAHAN, C. 1988. Marathon Oil Company internal report.

HOLMAN & W. E., ROBERTSON, S. S. 1994. Field Development, depositional model, and production performance of the turbiditic "J" sands at Prospect Bullwinkle, Green Canyon 65 field, outer-shelf Gulf of Mexico. *In*: Weimer, P., Bouma, A. H., Perkins, B. F. (eds.) Submarine fans and turbidite systems. GCSSEPM Foundation 15th Annual Research Conference, 139-150.

MAHAFFIE, M. J. 1994. Reservoir classification for turbidite intervals at the Mars discovery, Mississippi Canyon 807, Gulf of Mexico. *In*: Weimer, P., Bouma, A. H., Perkins, B. F. (eds.) Submarine fans and turbidite systems. GCSSEPM Foundation 15th Annual Research Conference, 233-244.

SHELTON, G. & HILL, W. A. 1993. Productive low resistivity well logs of the offshore Gulf of Mexico: a summary. *In*: Moore, D. (ed.), Productive Low Resistivity Well Logs of the Offshore Gulf of Mexico. Houston Geological Society and New Orleans Geological Society.

STEINMETZ, J. 1988. Marathon Oil Company internal report.

STOW, D. A. V. & PIPER, D. J. W. 1984. Deep-water fine-grain sediments: facies models. *In*: Stow, D.A.V. & Piper, D.J.W. (eds.), Fine-grain Sediments, Deep-water Processes and Facies. Blackwell Scientific publication, 611-647.

WEST, D. B. 1989. Model for salt deformation on deep margin of central Gulf of Mexico, A.A.P.G. Bulletin, v. 74, p. 1472-82.

GRAVITY-FLOW DEPOSITS OF A SLOPE VALLEY SYSTEM, HIGH ISLAND A-537 FIELD, GULF OF MEXICO

JOHN M. ARMENTROUT
Mobil Exploration and Producing Technical Center
PO Box 650232, Dallas, Texas 75265-0232

ABSTRACT

Conventional cores from four wells in the High Island A-537 field were examined, described, and sampled for sedimentologic, biostratigraphic, and kerogen analyses. The cores were recovered from the reservoir interval between the *Discoaster surculus* and the *Discoaster tamalis* Condensed Sections. 531 feet of core was examined of which 126 feet was sandstone, the remainder being siltstone and mudstone. Sedimentary features included massive sandstone, graded very fine-grained sandstone, laminated silty sandstone, laminated silty mudstone, massive mudstone, siltstone and mudstone couplets, rip-up clasts floating within the sandstone, convolute laminations, and climbing ripples. Sandstone and siltstone contain inner to middle neritic foraminifera while encasing mudstones have middle bathyal foraminiferal assemblages. Isolated kerogen analysis showed that over 80% of the kerogen was of terrestrial, woody origin. Paleogeographic reconstruction of the reservoir interval suggests deposition within in a salt-bounded submarine valley thirty-five miles from the age equivalent shelf-edge. All stratigraphic and sedimentologic data indicate that these sediments were deposited during a relative sea level lowstand by gravity-flow processes in a Lower Pleistocene to Upper Pliocene intraslope valley system.

INTRODUCTION

The High Island A-537 field is located in the High Island South Addition Area in the southwestern portion of the Plio-Pleistocene basin in the Texas Offshore (Fig. 1). The three blocks now included in the field were leased in OCS Sale 34 in 1974. The discovery well was drilled in Block A-537 with later drilling expanding the field into Blocks A-536 and A-531, along with Block A-556, acquired in 1986. Amoco, Mobil and Union of California were the operators of the field until its sale in 1993. Taylor and Armentrout (1990) published a paper on the geochemistry of the rocks and oils from the A-537 field, and much of this paper is taken from that study.

The reservoir interval occurs within a high-amplitude seismic package located on a northwest-southeast trending structural high which is flanked by two salt-withdrawal deeps to the southwest and northeast (Fig. 2). Isochrons of the overlying section suggest that the flanking depositional thicks filling the salt-withdrawal basins are younger than the high-amplitude reservoir package. This pattern of sediment accumulation encourages interpretation of the reservoir package as deposited within a regional low that subsequently became perched on a structural high formed by differential loading of salt by sediment..

The field has been developed from three platforms with a total of 40 wells (Fig. 3). Development of the field has been complicated by the presence of a shipping fairway down the middle of the accumulation, forcing every well in the field to be deviated. Production in the field began in September of 1978 with more than 140 MMBO and 110 BCFG having been recovered to date. The field reservoirs consist of six sandstones, designated BP-2, TP-

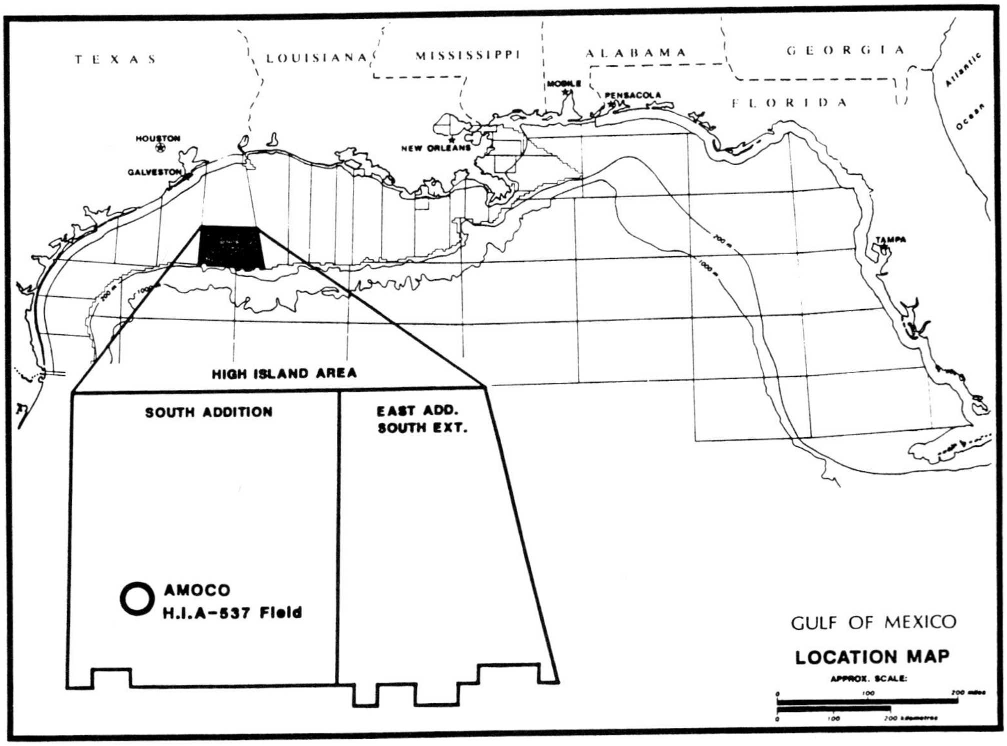

FIG. 1.—Location map showing the position of the High Island A-537 field.

1 through TP-4, and a "stray" sandstone. Only the TP-3 and TP-4 sandstones were cored by these wells and are discussed here. The cores examined were conventional 2.5 inch rubber sleeve cores and include the following intervals:

Block	Well	Core	Formation	Interval (MD in ft)
A-531	A-3	Amoco OCS-G-2696	TP-3 and TP-4	7360-7456
A-531	A-13	Amoco OCS-G-2696	TP-4	7601-7747
A-536	A-9	Amoco-OCS-G-2697	Shale	7612-7772
A-537	C-8	Amoco-OCS-G-2698	TP-3	9295-9424

The A-3 core is predominantly mudstone with intervals of sandstone which have either disaggregated or are missing (presumed lost during coring). The A-13 core was still in good condition having been kept frozen. The A-9 core is mudstone with a few thin poorly preserved sandstones. The C-8 core is almost entirely of poorly consolidated sandstone but having been kept frozen was in excellent condition when examined.

The sediments of the High Island A-537 field were deposited just prior to the extinction point of the benthic foraminifera *Textularia (N),* and between the fossil-rich condensed sections of *Discoaster surculus* and *Discoaster tamalis* (Schaffer, 1990). The approximate absolute age of this interval is 2.55 to 2.80 Ma (Fig. 4).

SEDIMENTATION AND STRATIGRAPHY

The cores consist of poorly indurated very fine-grained sandstone and mudstone. The reservoir sandstones are poorly consolidated, very friable and would best be described as very fine-grained, moderately sorted, submature subarkose (M.H. Link, 1987, Mobil internal report). They

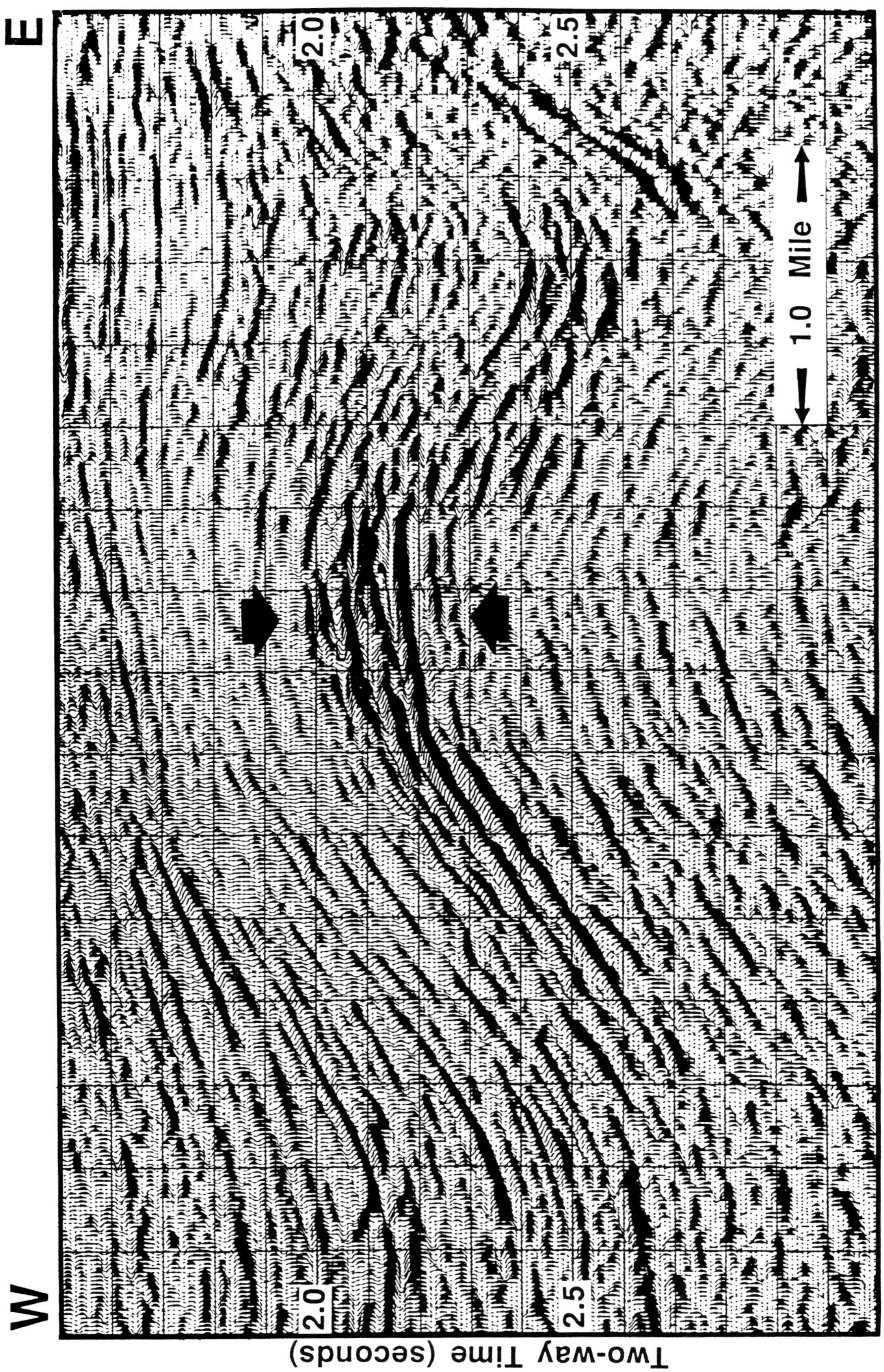

FIG. 2.—Seismic reflection profile showing the high-amplitude package associated with the High Island A-537 field reservoirs (between the arrows). Location of profile shown on Figure 3. Seismic used with permission of TGS-Calibre Geophysical Company.

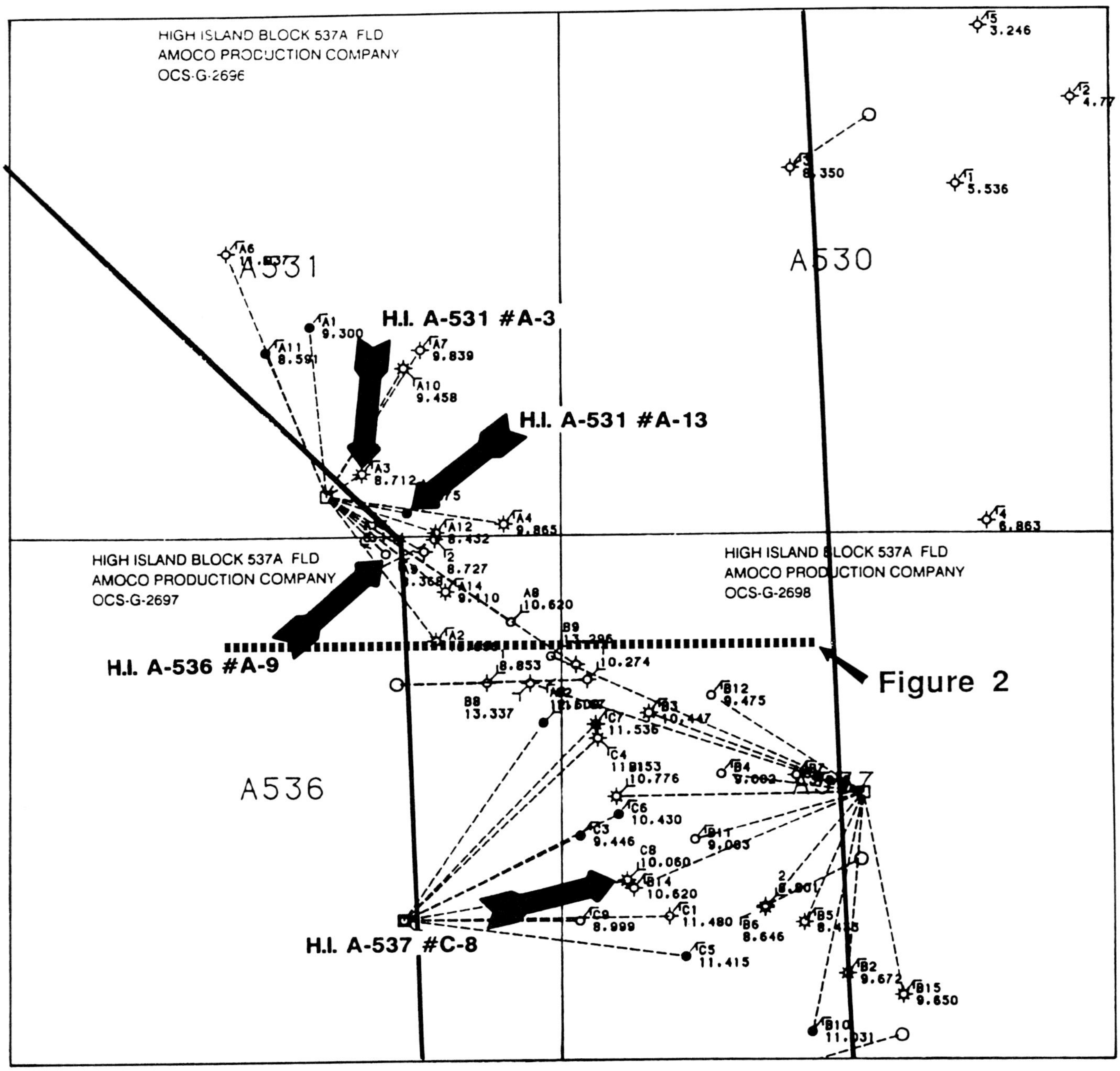

FIG. 3.—Field map, High Island A-537 field. Conventional cores used in this study came from the wells indicated by the large arrows. Reproduced from Taylor and Armentrout (1990).

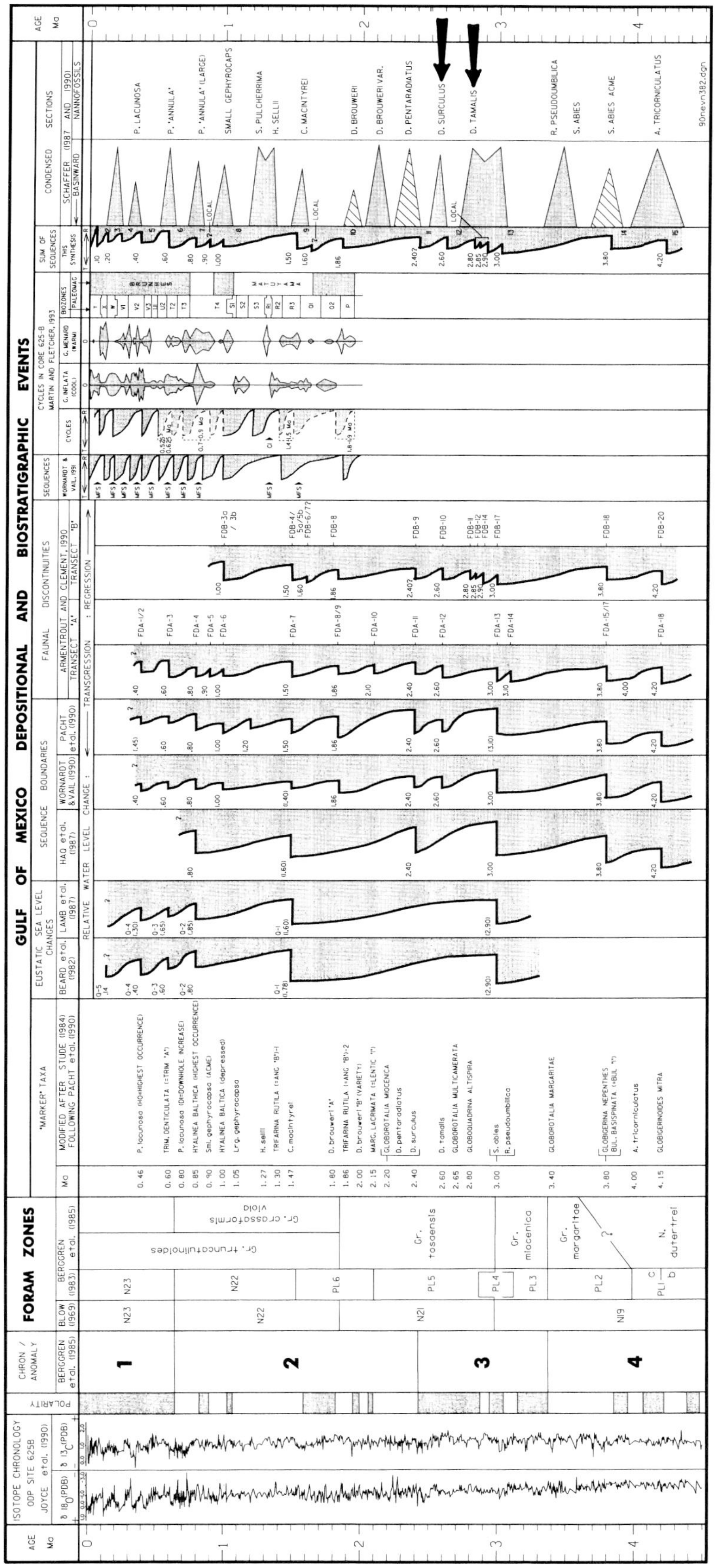

FIG. 4.—Stratigraphic correlation chart of Gulf of Mexico late Neogene depositional cycles. The two arrows at the right point at the two condensed sections bounding the High Island A-537 reservoir interval. The cycle chart is from Armentrout, in press.

have porosities in the 25 to 35 percent range; their permeabilities vary from the low hundreds of millidarcies to several darcies. Electric-logs of the cored intervals show predominantly blocky patterns for the sandstone intervals (Figs. 5 and 7).

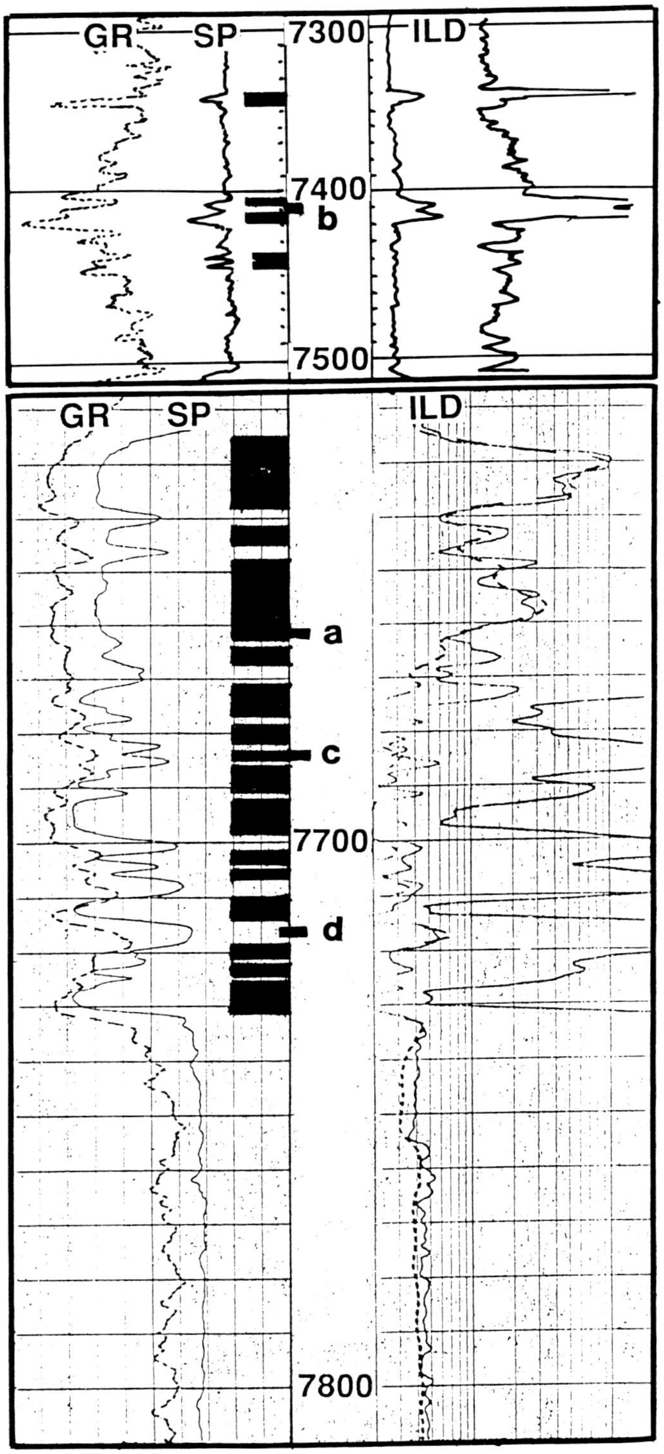

FIG. 5.—Electric-logs for the cored interval of the High Island A-531 A-13 well. Net-sand intervals are indicated by black boxes. Core photographs of Figure 6 are indicated by a, b, c, and d.

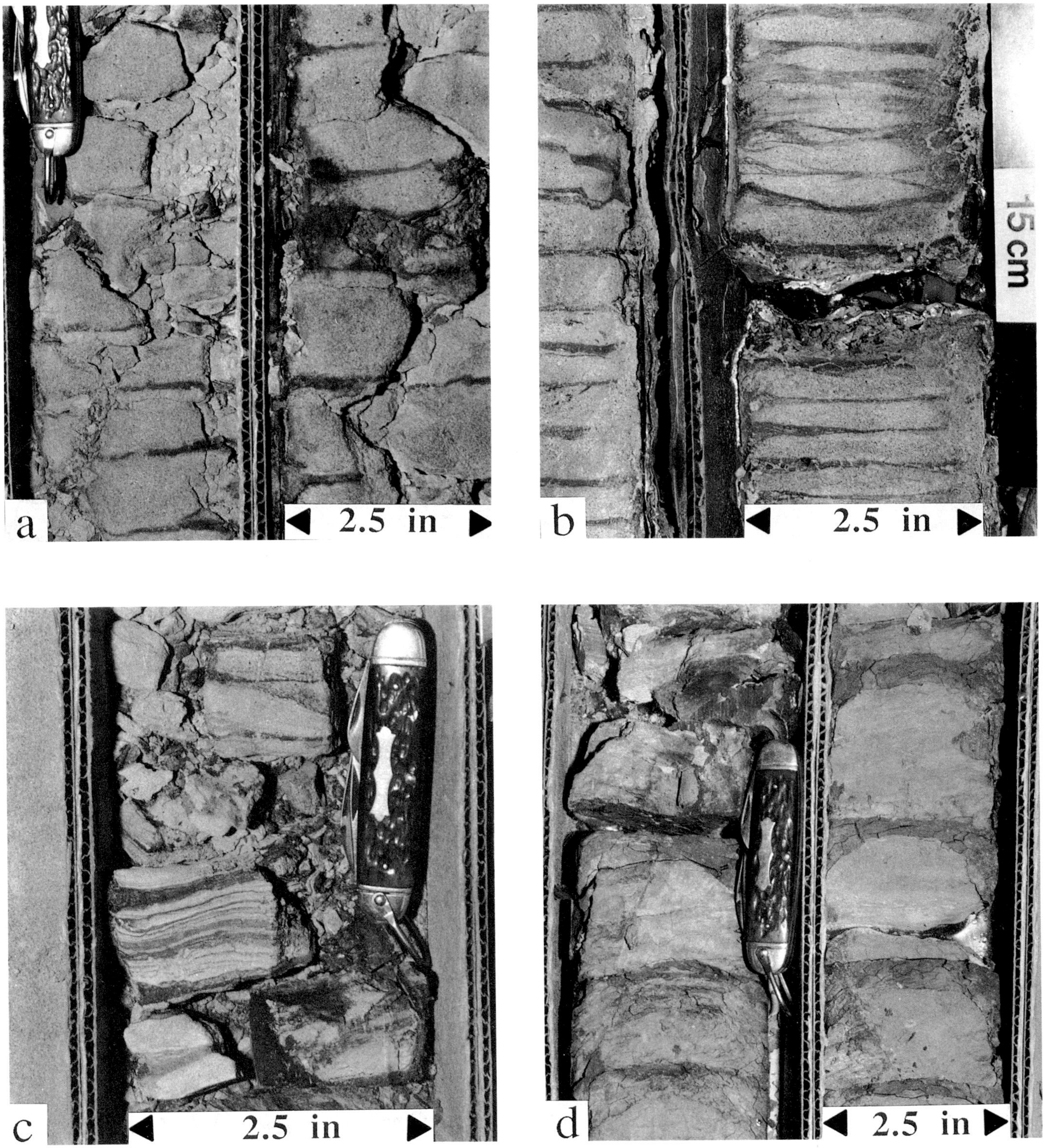

FIG. 6.—Core photographs of the High Island A-531 A-13 core. a. Graded beds of very fine-grained sandstone (7661 ft). b. Rippled sandstone lenses within mudstone (7416 ft). c. Laminated silty sandstones (7684 ft). d. Interbedded siltstone and dark mudstone (7718 ft). Photographs provided by P. Braithwaite.

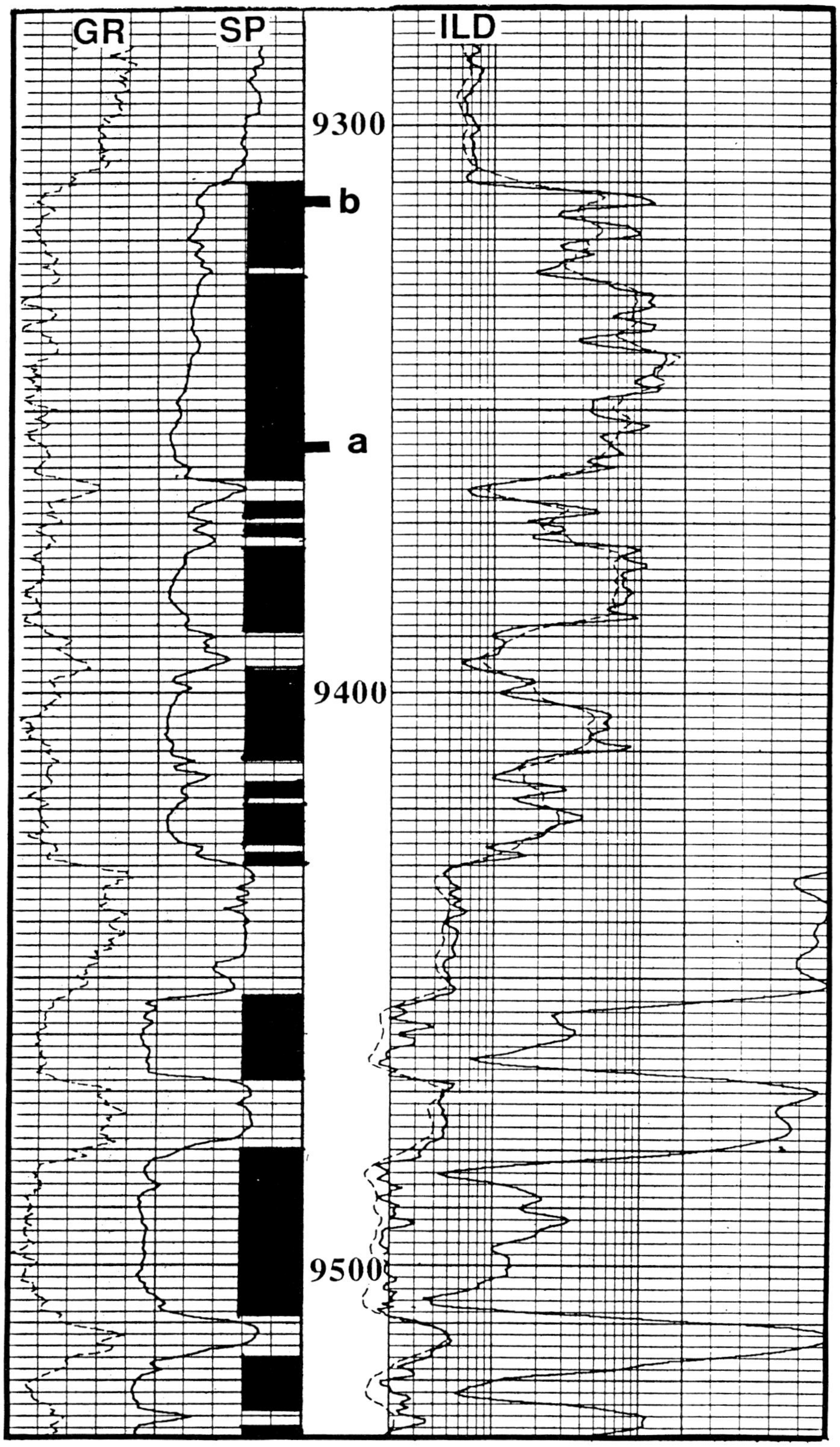

FIG. 7.—Electric-logs for the cored interval of the High Island A-537 C-8 well. Net-sand intervals are indicated by black boxes. Core photographs of Figure 8 are indicated by a and b.

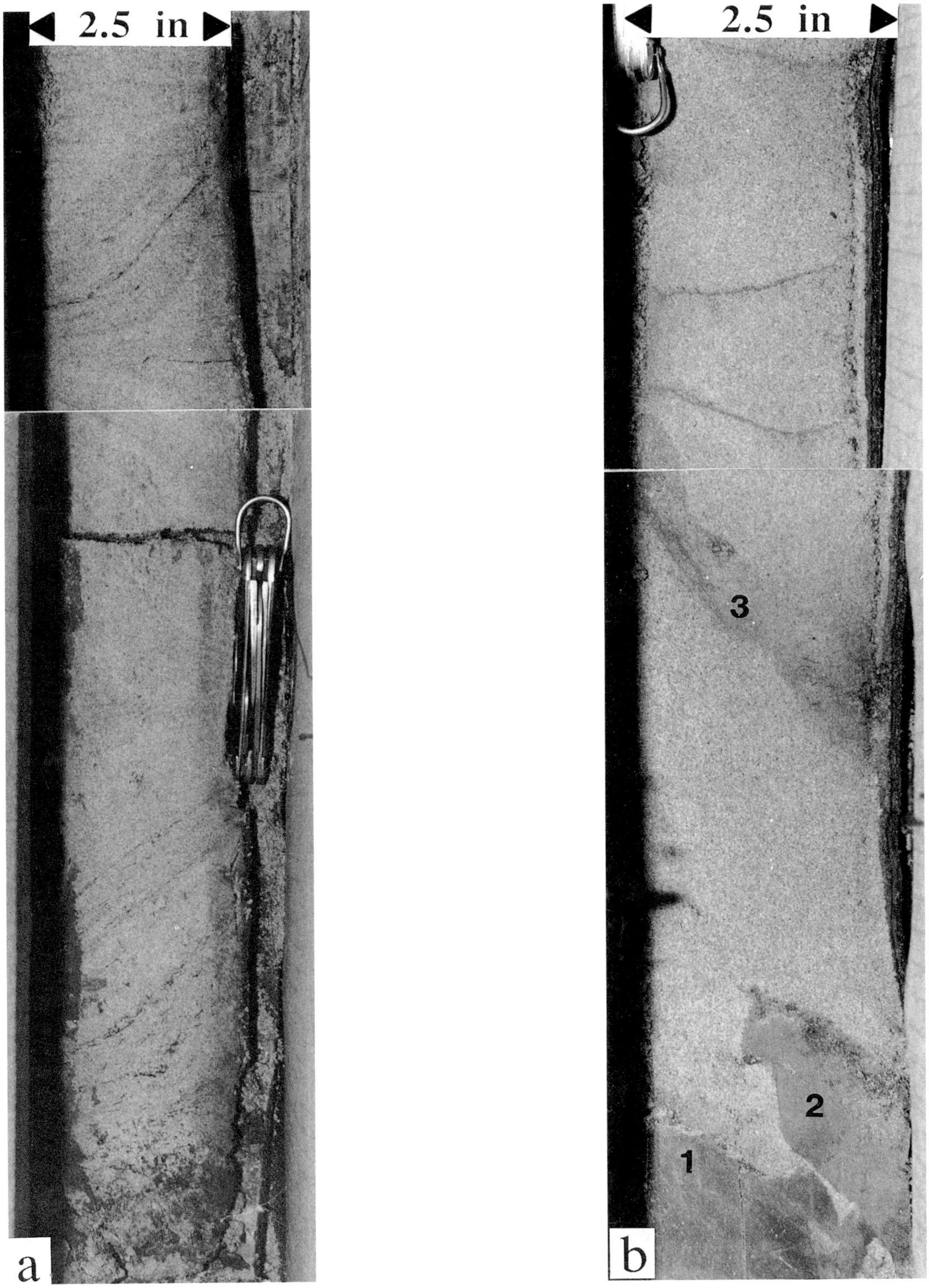

FIG. 8.—Core photographs of the High Island A-537 C-8 core. a. Convolute laminations (9356 ft). b. Base of channel eroded into mudstone (1) with rip-up clasts (2) and possible reactivation surface (3) (9313 ft). Photographs provided by P. Braithwaite.

The sandstones in the cores varied from thin beds (Fig. 6) to more massive units several feet in thickness (Fig. 8). The sandstones include one/half to two inch thick graded beds with relatively sharp bases and gradational tops (Fig. 6a), are commonly ripple laminated (Fig. 6b) with mudstone rip-up clasts, or are thinly laminated (Fig. 6c). The thicker sandstones tend to be more featureless and massive, although they sometimes exhibit convolute laminations (Fig. 8a) and rip-up clasts (Fig. 8b). The mudstones vary in thickness from thin laminae to massive, nondescript units several feet thick. Of particular note are couplets of siltstone overlain by mudstone forming rhythmically bedded intervals (Fig. 6d). Some mudstone layers contain wood fragments and mica flakes. Little of the mudstone appeared to be bioturbated, although few primary sedimentary structures were observed.

The reservoir sandstones contain rip-up clasts, sole marks, displaced shallow water marine fossils, laminated to massive appearing sandstones, internal scour features, all characteristics commonly associated with gravity-flow sandstones. The sandstones also show evidence of traction reworking in the form of climbing ripple marks and abundant ripple-laminated intervals. The sandstones can be grouped with facies A through G following the classification of Mutti and Ricci Lucchi (1972).

Paleoecological data from 109 samples taken for paleontological analysis clearly shows that the shales encasing the pay sandstones were deposited in a middle bathyal (1500-3000 ft water depth) environment (G.M. Ragan, Mobil Exploration and Producing United States, Personal Communication). Samples from the sandstones are strongly diluted by the high sedimentation rate and typically show an inner to middle neritic fauna. This shallow fauna is interpreted to be transported down-slope by gravity-flow processes and deposited along with the reservoir sands in middle bathyal depths.

The middle bathyal depositional setting for these sediments combined with the observed sedimentary structures and grain size distribution suggest that these sediments were deposited by gravity-flow processes including both debris flows and turbidity flows. The thick sandstones of core A-537 C-8 (Fig. 8) show both convolute laminations and rip up clasts suggesting plastic deformation, possibly by debris-flow events (see Shanmugam and others, this volume). The slightly graded sandstone/siltstone couplets of core A-531 A-13 (Fig. 6a) suggest moderately fluidized flow, possibly turbidites. The absence of Bouma sequences is attributed to the relatively uniform grain size of the sand fraction.

Total organic carbon measurements were made on 84 samples representing all lithofacies observed within the cores. The measured total organic carbon content (TOC) ranged from 0.03 to 0.94% with a mean of 0.36% and a standard deviation of 0.16% (Taylor and Armentrout, 1990). Without exception, the extremely low TOC values occurred in samples described lithologically as sandstone while the more organically enriched samples were described as mudstone.

The ten mudstone samples with the highest TOC values were used for isolated kerogen examination. In most samples woody plant debris comprised over 80% of the total kerogen. The bulk of the remaining kerogen was classed in the exinite fraction (Taylor and Armentrout, 1990). This mixture of kerogen best fits the TYPE III kerogen of Tissot et al., (1974), indicative of terrestrial origin.

DEPOSITIONAL SETTING

The depositional setting of the High Island A-537 field reservoir interval is interpreted to be an intraslope basin, trending northwest-southeast, in 1500 to 3000 or 4000 feet of water. This setting is illustrated on Figure 9, which combines a net-sand map, biofacies boundaries, and a seismically constrained shelf-edge. The net-sand map is constructed from more than 84 wells with the net-sand values contoured parallel to the thickness axes of the *Discoaster securlus - Discoaster tamalis* isochron. Foraminiferal biofacies boundaries provide approximately isobaths. The shelf-edge is identified by the topset/foreset inflection on seismic reflection profiles. The juxtaposition of the shelf-edge and middle-to-outer neritic biofacies boundary indicates a relative downward shift of environments from the modern highstand, suggesting that the *Discoaster securlus - Discoaster tamalis* isochron interval is a lowstand deposit. This lowstand slope deposit correlates with the lowstand between 2.8 and 2.7 Ma of Figure 4.

The A-537 field depositional basin was most likely formed as a result of salt withdrawal into salt ridges that flank the basin on both sides. This basin, approximately six miles wide and thirty miles long is best described as a valley. The valley opens into a large basin to the southeast which is not constrained by salt and is approximately twenty miles wide by at least thirty miles long, the southern margin being beyond current well control. The A-537 valley appears to be a narrow sediment pathway through which upper slope gravity-flow events transported sedi-

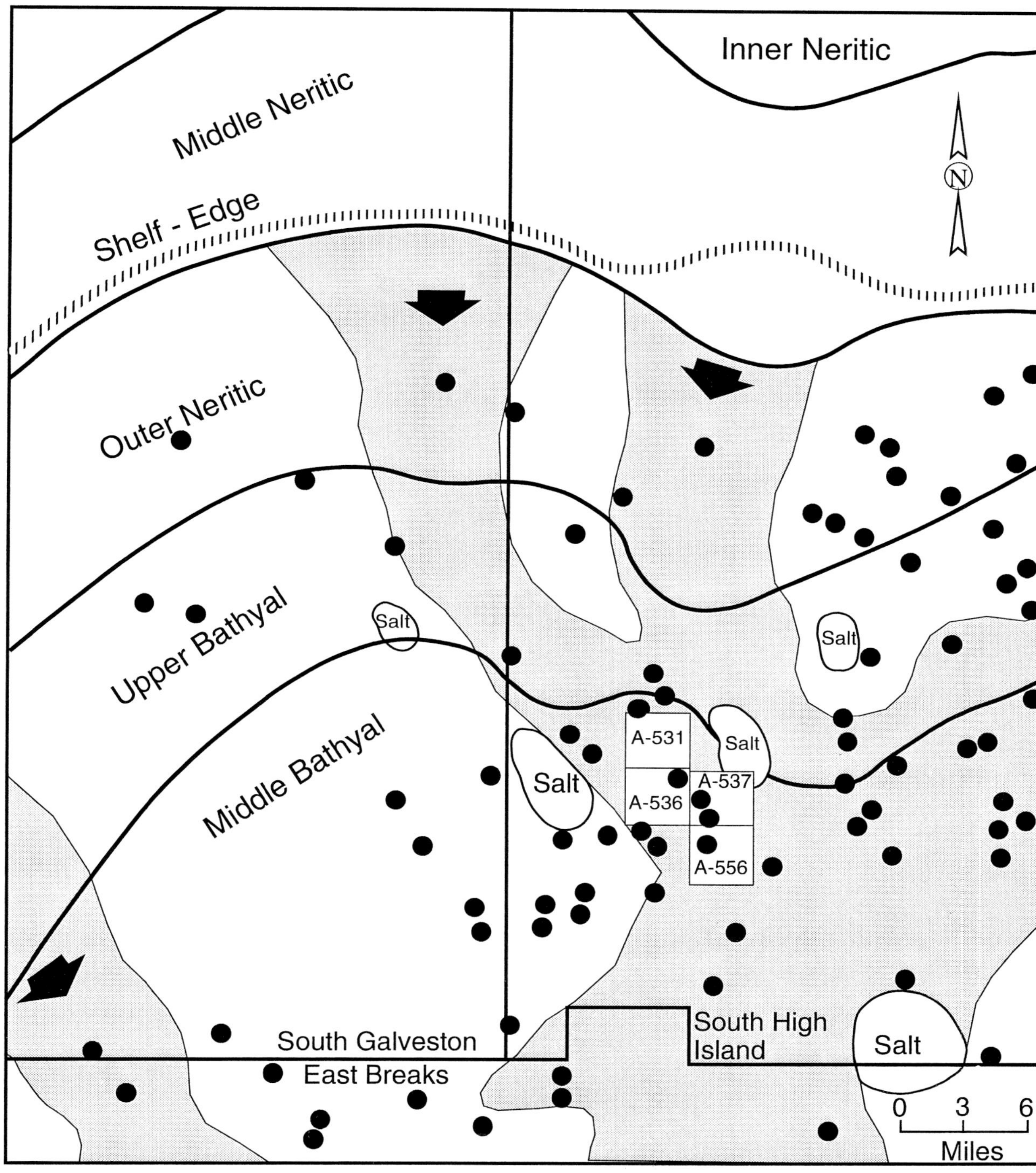

FIG. 9.—Net-sand map of the interval bounded by the *D. surculus* and *D. tamalis* Condensed Sections (see Fig. 3). Map area shown for net-sand greater than 25 feet (stippled). Block-outline shows location of A-537 field. Dots indicate well control for net-sand map. Heavy black lines are boundaries of foraminiferal biofacies. Salt shown where it interrupts the mapped interval. Broken line shows approximate shelf-edge interpreted from topset-foreset inflection on seismic reflection profiles. Net-sand map modified from unpublished map by T.D. Crutcher, 1986, Mobil internal report.

...e more distal and less geographically con-
...d basin.

Multiple gravity-flow events probably passed through this valley forming long narrow channel systems in the physiographic low. Subsequent events filled the channels with geographically restricted channel-fill and channel overbank deposits. These sediments are restricted in their aerial extent to the valley axis, probably anastomosing in pattern, and stacked consecutively as a function of valley floor topography. This model in agreement with the depositional model proposed for the High Island A-560-561 field located eight to ten miles southwest (Lund et al., 1978), and with the model proposed for the East Breaks 160-161 field located eighteen miles to the west southwest (Armentrout et al., 1991).

ACKNOWLEDGMENTS

The author gratefully acknowledges the contributions of several co-workers for elements of this paper. The original work was initiated by Gilbert D. Taylor; Marty H. Link made the original core descriptions; Philip Braithwaite took the photographs; Gerry M. Ragan made all biostratigraphic analyses; and Tom D. Crutcher provided the regional net-sand map. Amoco Production Company has given permission to use the core; TGS-Calibre Geophysical Company allowed use of the seismic reflection profile; and Mobil Oil Corporation gave permission to publish the paper. The manuscript has been improved by reviews provided by Steve M. Mitchell and Kathy A. Kanschat.

REFERENCES CITED

Armentrout, J.M., S.J. Malecek, P. Braithwaite, and C.E. Beeman, 1991, Seismic facies of slope basin turbidite reservoirs, East Breaks 160-161 field: Pliocene-Pleistocene, northwestern Gulf of Mexico. *In*: Weimer, P. and M.H. Link, (eds.), Seismic Facies and Sedimentary Processes of Submarine Fans and Turbidite Systems: Springer-Verlag, New York, p. 223-239.

Armentrout, J.M., (in press), High-resolution sequence biostratigraphy: Examples from the Gulf of Mexico Plio-Pleistocene: High-Resolution Sequence Stratigraphy, The Geological Society, London (1995).

Lund, J.W., J.S. King, R. Berlitz, and J.A. Gilreath, 1978, Pre-platform exploration of High Island Blocks A-560 and A-561: Gulf Coast Association of Geological Societies Transactions, v. 27, p. 273-294.

Mutti, E., and F. Ricci Lucchi, 1972, Turbidites of the northern Apennines; Introduction to facies analysis: International Geological Reviews, v. 20, p. 125-166 (English translation by T.H. Nilsen).

Schaffer, B.L., 1990, The nature and significance of condensed sections in Gulf Coast late Neogene sequence stratigraphy: Gulf Coast Association of Geological Societies Transactions, v. XL, p. 186-195.

Taylor, G.D., and J.M. Armentrout, 1990, Rock geochemistry and relationships to produced oils from upper Pliocene turbidites, High Island Area, Gulf of Mexico. *In*: Schumacher, D., and B.F. Perkins, (eds.), Gulf Coast Oils and Gases; Their Characteristics, Origin, Distribution, and Exploration and Production Significance: Gulf Coast Section Society of Economic Paleontologists and Mineralogists Foundation Ninth Annual Research Conference Proceedings, p. 151-161.

Tissot, B., B. Durand, J. Espitalie, and A. Combaz, 1974, Influence of nature and diagenesis of organic matter in formation of petroleum: American Association of Petroleum Geologists Bulletin, v. 58(3) p. 499-506.

CARDIFF
UWCC LIBRARY